# US-Straßenkreuzer

- Station-Wagons
- Corvette

Schrader-Motor-Chronik

Motor buch Verlag

Einbandgestaltung: Luis dos Santos



ISBN 978-3-613-03392-4



Sie finden uns im Internet unter:
www.motorbuch-verlag.de

1. Auflage 2012



Druck und Bindung: KoKo Produktionsservice, 70900 Ostrava
Printed Czech Republic

Schrader-Motor-Chronik

# CHEVROLET CORVETTE
## 1953–86

Dokumentation: Walter Zeichner

Als den einzig echten Sportwagen Amerikas apostrophierte General Motors den rassigen Zweisitzer bei seiner Vorstellung im Jahre 1953. In der Tat gehörte dieser Wagen zum Faszinierendsten, was Detroit je zu bieten hatte. Ein Querschnitt durch die Fülle der mit Liebe ausgewählten und in diesem Buch wiedergegebenen Werbemittel und Presseberichte dreier Jahrzehnte macht deutlich, warum die Corvette so erfolgreich war. Denn dieses Auto war alles andere als der „typische Amischlitten" – die Corvette galt stets als ein Fahrzeug mit eigenem Charakter.
Die Zusammenstellung der Dokumente besorgte Walter Zeichner mit Unterstützung durch Kai Jacobsen, dem hierfür besonderer Dank ausgesprochen sei.

Halwart Schrader
Herausgeber

## Inhalt

## Ein Sportwagen made in U.S.A

Die Corvette war und ist auch heute der einzige echte Sportwagen aus amerikanischer Produktion, sieht man einmal von ein paar kurzlebigen ,,Eintagsfliegen'' ab. Es erstaunt diese Tatsache um so mehr, denkt man an den hohen Stellenwert, den der Motorsport in den USA seit jeher einnimmt.
Dabei bestand zwar schon immer ein sehr hoher Bedarf an Sportwagen, der aber fast ausschließlich durch Fabrikate europäischer Herkunft gedeckt wurde.
Die Entstehungsgeschichte der Chevrolet Corvette ist eng verknüpft mit dem Designer Harley Earl, der schon 1951 die Idee zu einem preiswerten, einfach herzustellenden Sportwagen gehabt hatte und der es schließlich durchsetzen konnte, daß solch ein Wagen auch tatsächlich gebaut wurde.
Earl war 1932, nach dem Studium in Stanford und einer Zeit, in der er Spezialkarosserien für eine extravagante Clientel in Hollywood gezeichnet hatte, zu General Motors gestoßen. Er hatte dort ein Design Center aufgebaut, das damals seinesgleichen suchte und in dem Leute arbeiteten, die den Stil der amerikanischen Automobile maßgebend beeinflußten; Maßstäbe setzten dabei die anläßlich von Shows und Motoramas gezeigten ,,Dream Cars'', Wagen mit auffällig futuristischem Styling und verspielten Details, die einen Blick in die Zukunft des Automobils, jedenfalls in den USA, gewähren ließen und um die sich ein autobegeistertes Publikum drängte. Solche Schöpfungen waren eine Spezialität Harvey Earls. Mit seinem 1938 vorgestellten ,,Buick Y-Job'' beeinflußte er das Ideal amerikanischer Autobaukunst bis in die späten vierziger Jahre.
Nach dem zweiten Weltkrieg, nachdem die amerikanischen Automobilgiganten voll mit der Produktion von Kriegsmaterial beschäftigt gewesen waren und die Herstellung von Zivilfahrzeugen geruht hatte, wurde erst einmal mit den alten Modellen die Produktion fortgesetzt, bevor mit Beginn der fünfziger Jahre bei allen amerikanischen Automobil-Herstellern neue Stilelemente sichtbar wurden.
Einer dieser Trendsetter war der von Earl entworfene Buick ,,Le Sabre'', ein vom Düsenjäger-Styling beeinflußter Prototyp, der 1951 auf den Shows für Furore sorgte und von dem sich einige markante Merkmale, wie die extremen Panoramascheiben und das unter einem Deckel verschwindende Faltdach, auf die spätere Corvette vererben sollten.
Im gleichen Jahr hatte Harvey Earl mit einem kleinen Stab persönlicher Mitarbeiter in einem separaten Studio mit der Entwicklung seines Sportwagens begonnen. Zuerst war geplant, einen Zweisitzer zu bauen, der in der Preislage normaler Chevrolet- oder Ford-Limousinen rangierte. Erste Zeichnungen zeigten ein Auto, das wie eine Kreuzung zwischen Jaguar XK 120 und Jeepster aussah und auf einem normalen Chassis der laufenden Limousinenreihe basierte. Die Arbeiten liefen unter der Bezeichnung ,,Projekt Opel'', denn man legte auf strenge Geheimhaltung auch innerhalb des Konzerns Wert, der für die deutsche Tochter schon öfter Design-Arbeiten übernommen hatte.
1952 stieß mit Edward Cole ein Konstrukteur zum Team, der die technische Seite der künftigen Corvette prägen sollte; bald arbeiteten beide Abteilungen eng zusammen. Earls Mannschaft kam immer weiter von der ersten nüchternen Planung ab und orientierte sich stilmäßig nun stark an Dream Cars wie Buick Le Sabre und dem neuen Typ XP-300, einem noch futuristischeren Modell.
Es wurde indessen vorausgesetzt, daß der Wagen mit dem geringsten Aufwand an neu zu konstruierenden Teilen herzustellen war, denn in den Chefetagen wäre ein kostspieliges ,,Spielzeug'', von dem keine sehr hohen Stückzahlen zu erwarten waren, sonst auf taube Ohren gestoßen. So mußte in dem Sinne konstruiert werden, daß möglichst viele Teile ,,aus dem Regal'' der aktuellen

*Da freut sich die Frau Gemahlin des Generaldirektors – so ungefähr konnte die Bildaussage dieses Pressefotos aus dem Jahre 1960 lauten. Zentralverschluß-Attrappen und Weißwandreifen waren stets beliebt.*

Produktionsteile zu nehmen waren, was so gut wie alle mechanischen Teile und das Chassis betraf. Auch die Karosserie sollte so preisgünstig wie möglich zu bauen sein.
Nun hatte man bei der Chevrolet Division gerade ein großes Cabriolet mit verstärkter Fiberglas-Karosserie ge-

*Das 1964er Sting Ray Coupé. War Fliegen schöner? Der Flugkapitän und die Stewardeß hätten zugunsten der Corvette gestimmt . . .*

testet. Das neue Material hatte sogar einen unfreiwilligen Überschlag des Testfahrers fast unbeschädigt überstanden. Harley Earl kam dies wie gerufen, denn durch dieses leicht zu formende, neue Material waren enorme Kosten für Karosseriewerkzeuge einzusparen, womit sein ,,Baby" ein noch geringeres Investitionsrisiko für das Unternehmen darstellen würde. Die Chancen für grünes Licht zur Serienproduktion stiegen; wie sich das Kunststoffmaterial dann im jahrelangen Alltagsbetrieb benehmen würde, war eine andere Frage.

Inzwischen war es Sommer 1952 geworden, und ein fertiges Tonmodell des neuen Zweisitzers stand in der Entwicklungsabteilung des Art & Color Studios. Auch Chevrolet-Chefingenieur Ed Cole war begeistert von dem Wagen. Nun kam der große Moment – die Präsentation des Wagens vor General-Motors-Präsident Harlow Curtice und der Chevrolet-Vorstandschaft. Der neue Sportwagen gefiel den hohen Herren so gut, daß ein Prototyp in Auftrag gegeben wurde, um ihn anläßlich des ersten Motoramas der Öffentlichkeit zu präsentieren. Man versprach sich durch solch einen Wagen vor allem auch eine Imageverbesserung der ganzen Chevrolet-Linie.

Auf Hochtouren wurde jetzt an der Fertigstellung des Prototyps gearbeitet, der ja auch unter der Haube voll funktionsfähig sein mußte. Fahrwerkspezialist Maurice Olley schaffte es, ein passendes Chassis und eine Radaufhängung weitgehend aus Serienteilen zu bauen, die

nur leicht für die Anforderungen eines Sportwagens modifiziert waren. Das wirkte sich positiv auf die Kosten aus, wenngleich der Wagen in konstruktiver Hinsicht mit den europäischen Produkten nicht mithalten konnte.
Am 17. Januar 1953 wurde schließlich der Schleier gelüftet. Im großen Ballsaal des New Yorker Waldorf-Astoria-Hotels drängten sich die Besucher der ersten Motorama des Jahres vor dem weißen Roadster von Chevrolet. Der Radstand des „Corvette" genannten Zweisitzers entsprach genau dem des Jaguar XK 120, doch war der Wagen breiter und wuchtiger. Man hatte es deutlich vermieden, Stylinganleihen aus dem europäischen Sportwagenbau zu übernehmen, und ein typisch amerikanisches Auto geschaffen. Um die Gewichtsverteilung zu optimieren, lagen die Sitze direkt vor der Hinterachse. Der Motor, ein frisierter „Blue Flame" Sechszylinder mit 150 PS Leistung, konnte weit hinten installiert werden, so daß sich das Gewicht 53 zu 47 auf Vorder- und Hinterachse verteilen ließ. Ein Kühlermaul mit großen Chromzähnen beherrschte die Front, die nur von Stoßstangenhörnern rechts und links des Nummernschildes und an den Ecken geschützt wurde; sportliche Steinschlaggitter zierten die ovalen, tief in die runden Vorderkotflügel integrierten Scheinwerfer.
Eine niedrige Panoramascheibe begrenzte den mit rotem Kunstleder ausgeschlagenen Innenraum, deren Chromrahmen sich in einen das ganze Cockpit umfassenden Ring fortsetzte. Auch das Armaturenbrett erinnerte sehr an Dream Cars. Links war der Tacho eingebaut, während rechts Lautsprecher und Ablagefach untergebracht waren. Dazwischen schauten den Passagieren nicht weniger als sechs Anzeigeinstrumente entgegen, das größte davon in der Mitte war der Drehzahlmesser. Die Schalensitze waren eingebettet in die Karosserie, und die Türen des Prototyps öffneten sich auf Knopfdruck. Das leichte Verdeck mit Steckscheiben verschwand unter einem Deckel hinter den Sitzen. Das Heck lief rund aus wie die Front, nur waren hier deutliche Flossen einmodelliert worden mit runden Heckleuchten an der Spitze und kleinen Torpedoflügeln. Die großen Radausschnitte gaben freien Blick auf die Weißwandreifen, und eine von den vorderen zu den hinteren Stoßfängern laufende seitliche Zierleiste flankierte die großen Seitenflächen.
Es war ohne Zweifel ein sehr gelungenes und gefälliges Auto, das die Leute in New York bestaunen konnten. Fast vier Millionen begutachteten den neuen Typ. Die Reaktionen waren überwältigend. Harley Earl hatte seinen Traum verwirklicht – die Firmenleitung gab die Anweisung, die Corvette so schnell wie möglich in Produktion gehen zu lassen.
Noch im selben Jahr 1953 entstanden die ersten 300 Vorserienexemplare weitgehend in Handarbeit, während man in St. Louis eine neue Produktionsstraße einrichtete, die dann bis 1981 Geburtsort der Corvette bleiben sollte. Damals wurden die 46 Karosserieteile aus Glasfiber von einer Fremdfirma in Ohio gefertigt und dann von Hand zusammengeleimt, während die Motoren aus dem Chevrolet-Motorenwerk in New York stammten. Auch den Rahmenbau nahm eine Vertragsfirma vor. Diese ersten Wagen waren noch nicht perfekt in der Verarbeitung, und die ersten 25 hatten auch die Serienradkappen der Chevrolet „Bel Air" Limousine, bevor es eigene Radkappen für die Corvette mit Zentralverschluß-Attrappe gab (was viele indessen als wenig geschmackvoll empfanden). Alle 300 Wagen waren in Polo-Weiß gespritzt und hatten rot/weißes Interieur mit schwarzem Verdeck sowie Weißwandreifen.
Leider enttäuschte jedoch schon die ersten Tester und Eigentümer die Leistung des Wagens, der die hochgeschraubten Erwartungen keinesfalls zu erfüllen vermochte. Der Sechszylindermotor ließ den Wagen, der aus Kosten- und Platzgründen mit der bekannten Powerglide-Zweigangautomatik ausgerüstet war, nur

knapp 170 km/h erreichen. Auch eine Beschleunigungszeit von bestenfalls 11 Sekunden von 0 auf 100 km/h war nicht dazu geeignet, das verwöhnte amerikanische Publikum zu begeistern. Zugleich hatte man den geplant niedrigen Preis bei weitem nicht realisieren können, und so kostete der Roadster jetzt beachtliche 3513 Dollar.
Für das Modelljahr 1954 erfolgte dann die Produktion in St. Louis mit etwa 50 Wagen täglich. Im Laufe des Jahres erfuhr die Corvette dann schon einige Verbesserungen. Außer in Polo White konnte man den Sportwagen in Metallic-Blau haben, aber nur vier Prozent entschlossen sich für die dritte Möglichkeit, „Sportsman Red". Außerdem hatten von den 3640 Corvetten, die in diesem Jahr entstanden, nur sechs eine schwarze Sonderlackierung in Verbindung mit roter Innenausstattung.
Um die Türen der Serien-Corvette zu öffnen, gab es keinen Druckknopf mehr wie beim Prototyp, sondern man mußte – was vor allem bei geschlossenem Verdeck keine Freude bereitete – durchs Fenster greifen und den Innenhebel nach hinten ziehen. Verbessert wurde hingegen die Position des Choke, der bei den ersten Wagen noch rechts neben dem Anlaß-Schloß lag, so daß man mit der linken Hand durchs Lenkrad fassen mußte und gleichzeitig mit der rechten Hand den Zündschlüssel umdrehte. Das Verdeck präsentierte sich bei den 1954er Modellen braun.
Die Karosserie machte noch einige Sorgen, da es an manchen Stellen Wassereinbrüche zu verzeichnen gab, ein Mangel, der erst im Laufe dieses Modelljahres behoben werden konnte. Dennoch gab das Desinteresse des Publikums an der Corvette zu Sorgen Anlaß: Die Kritiker waren offenbar zu hart mit dem Auto umgegangen. Mitt lerweile an europäische Formen und Hochleistungsmotoren gewöhnt, ließ man kaum ein gutes Haar am Chevrolet. Für einen Sportwagen fehlte dem Auto nach Meinung vieler die Rasanz. Die Automatik und der biedere Motor machten zusammen mit der bauchigen Karosserie einem Jaguar keine Konkurrenz, und auf der anderen Seite war das Auto zu unkomfortabel für einen Tourer. Ende 1954 standen immer noch fast 1500 Wagen bei den Händlern auf dem Hof. Doch für das Modelljahr 1955 gab es gute Nachrichten, denn auf Wunsch wurde nun der neue, von Ed Cole konstruierte V8-Motor eingebaut, der 195 PS brachte und dabei noch 15 Kilo leichter war als der alte Sechszylinder. Diese erste V8 Corvette brauchte nur noch 8,5 Sekunden für den Spurt auf 100 km/h – über mangelnde Power konnte man sich also nicht mehr beschweren.
Das Automatik-Getriebe entsprach dem Image eines Sportwagens ebenfalls nicht, doch es blieb zunächst im Programm. In diesem Modelljahr fanden sich nur 700 Käufer, von denen ganze zehn sich noch für den Sechszylinder entschieden. Wer einen „richtigen" Sportwagen wollte, hatte schließlich eine große Auswahl an schicken Wagen aus Europa, und wer es amerikanisch-gemütlich, aber trotzdem mit sportlicher Aufmachung und guter Leistung liebte, der kaufte sich den neuen Ford Thunderbird, der limousinenhafte Bequemlichkeit mit preudo-sportlichen Attributen perfekt kombinierte und sich mehr als 20mal so gut verkaufen ließ wie die Corvette, deren Tage schon gezählt schienen.
Doch General Motors wollte das Rennen gegen die Ford Company unter keinen Umständen verlieren. Schließlich trat ein Mann auf den Plan, der für viele Jahre die Geschicke dieses amerikanischen Sportwagens lenken sollte: Zora Arkus-Duntov. Der in Belgien geborene ehemalige Konstrukteur bei der englischen Sportwagenfirma Allard war im Geburtsjahr der Corvette zu GM gestoßen und hatte sich bald des erfolglosen Roadsters angenommen. Das neue Modell 1956 sollte er bereits entscheidend prägen. Vor allem wurde die Karosserie komplett überarbeitet, wobei die vordere Hälfte erhebliche Veränderungen erfuhr. Eine breite Mulde war in die Flanken einmodelliert worden, die vom Radausschnitt

bis fast unter die neuen Türgriffe reichte und in kontrastierender Farbe gespritzt wurde. Die Scheinwerfer standen jetzt senkrecht und hatten keine Schutzgitter mehr. Am Heck waren die Flossen weggefallen, und übrig blieben schön geschwungene Kotflügel, an deren Enden die beiden Auspuffrohre mündeten, die beim Vormodell neben dem Kofferdeckel ins Freie geragt hatten und durch Verwirbelungen der Abgase stets für ein verschmutztes Heck des Roadsters gesorgt hatten. Für die komfortbewußten amerikanischen Sportfahrer hatte der Wagen nun auch Kurbelfenster, und auf Wunsch machte ein Hardtop die Corvette zum Coupé.
Der Wagen konnte jetzt in sieben Farben bestellt werden mit mehr oder weniger geschmackvollen Interieur-Varianten. Doch beschränkten sich die Veränderungen für die Corvette der zweiten Generation keineswegs auf Äußerlichkeiten. Der aktive Rennfahrer Arkus-Duntov hatte seine Aufmerksamkeit merklich den Fahreigenschaften gewidmet (nur die Bremsen waren leider immer noch weit überfordert und reagierten bei sportlicher Fahrweise schnell mit Fading). Vor allem aber hatte die Corvette jetzt eine Dreigang-Schaltung anstatt der Automatik bekommen, zusammen mit einem in der Leistung abermals angehobenen Motor. Der heißeste Limousinenmotor mit Vierfachvergaser war durch den Einbau einer von Duntov „verschärften" Nockenwelle auf 225 PS bei 5200/min getunt worden; mit dem Schaltgetriebe stürmte die neue Corvette nun in beachtlichen 7,5 Sekunden auf 100 km/h und dies sogar mit der langen Übersetzung von 3,27 : 1.
1958 gab es erneut ein deutliches Face-lifting mit Doppelscheinwerfern, zusätzlichen Lufteinlässen vorne und reichlich mehr Chromzierat; alles in allem wurde der Wagen dadurch fast einen Zentner schwerer, aber kaum schöner; er wirkte jetzt reichlich überladen. Innen gab es ein völlig neues Armaturenbrett in Form zweier halbrunder Einbuchtungen. Die rechte war mit einer dicken Haltestange versperrt, an der sich der Beifahrer bei kühnen Manövern festklammern konnte, und auf der Fahrerseite waren nun alle Anzeigeinstrumente außer der Uhr versammelt, überragt vom riesigen, halbrunden Tacho.
Als Motorisierung gab es bis 1961 nur noch das 283-c.i.-Aggregat mit Leistungen zwischen 230 (mit Vierfachvergaser) und 315 PS (mit Benzineinspritzung). Keine Corvette war jetzt länger als 8 Sekunden bis zur 100-km/h-Marke unterwegs.
1961 – schon unter der Regie des neuen Chefdesigners Bill Mitchell, der die futuristischen Dream Cars „Stingray" und „XP-700" entworfen hatte – wurde die runde Heckpartie durch ein „ducktail"-Heck ersetzt. Der Trend ging zu eckigeren Formen. Im folgenden Jahr wurden die Seitenflächen schlichter, außerdem gab es wieder einen größeren Achtzylinder mit nun 327 c.i. Mit Vergasern gaben diese Motoren zwischen 250 und 340 PS ab, und eine mit 11,25 : 1 enorm hoch verdichtete Version mit Benzineinspritzung produzierte sogar 360 PS. Ein solcher Wagen kostete schon knapp über 5000 Dollar.
Stilistisch stark an den „Stingray"-Prototyp angelehnt, erschien 1963 die vierte Generation der Corvette in Form von Roadster, Hardtop und einem neuen Coupé mit ungewöhnlich gestyltem, spitz zulaufendem Fließheck mit Doppelfenster und weit ins Dach hineinreichenden Türen – heute eine der gesuchtesten Corvette-Varianten. Bill Mitchell hatte den neuen Stil geprägt, mit Haifischnase und -maul und im Vergleich zum Vormodell kantigerer Linienführung. Doch auch unter der Kunststoffhaut hatte sich viel getan. Ein neuer Leiterrahmen machte es möglich, die Sitze noch tiefer und weiter hinten unterzubringen, um die Gewichtsverteilung zu optimieren. Der Radstand war verkürzt worden, und hinten hatte man die Räder nun endlich einzeln aufgehängt, was die Straßenlage des Autos sehr verbesserte.

Für das Modelljahr 1963 hatte man die Motoren vom Vorjahr übernommen, doch ab 1964 und bis zum Ende dieser Karosserievariante 1967 fand nochmals eine wahre Leistungsorgie statt. 1964 gab es sechs Leistungsversionen des 327-c.i.-Motors von 250 bis 375 PS, doch das war erst der harmlose Anfang. 1965 – im gleichen Jahr gab es erstmals auf Wunsch Scheibenbremsen an allen vier Rädern – kam ein neuer 396-c.i.-Motor mit 425 Vergaser-PS, der zusammen mit einem Sportfahrwerk und optischen Signalen wie dickem Buckel auf der Motorhaube und Sidepipes den Chevy Sportwagen in ein aufregendes Vehikel verwandelte.
Ab jenem Jahr gab es aus Kostengründen keine Einspritzmotoren mehr, doch die Corvette hatte sich mittlerweile zu einem Verkaufsrenner entwickelt. Fast 120 000 Wagen wurden allein von dieser Karosserieversion verkauft, davon etwas über 72 000 Roadster.
Schon 1963/64 hatte man mit den Arbeiten an einem Nachfolgemodell begonnen. Es handelte sich um ein wunderschönes Mittelmotor-Versuchsmodell mit der Bezeichnung ,,XP-882", das jedoch wenig später aus Kostengründen zu den Akten gelegt wurde. Die neue Corvette wurde weiterhin konventionell gebaut. Als Vorbild diente die schon fast bösartig aussehende Styling-Studie ,,Mako Shark II" von 1965, die unter der Regie des neuen Art-Directors David Holls zu einem etwas gemäßigteren, wenn auch immer noch äußerst aggressiv wirkenden Sportwagen geformt wurde.
Die Kritiker hatten stets viel zu bemängeln gehabt an der hierzulande vielleicht bekanntesten Corvette-Form. Sie war ihnen zu groß, zu altmodisch, und im Innenraum gab es zu wenig Platz für Passagiere und Gepäck. Trotzdem ließen sich im Modelljahr 1968 immerhin 28 566 Käufer nicht von einem Kauf abhalten – ein neuer Rekord. Bis 1983 wurde diese Karosserie mit zahlreichen Veränderungen gebaut. Es gab drei Varianten bis zum Wegfall des Roadsters 1976, nämlich zusätzlich ein ,,T-Top" Cabrio mit abnehmbaren Dachhälften sowie ein Coupé.
Scharfe Abgasgesetze drückten die Leistung der stärksten Version 1972 auf 365 PS, und nach einigen Zwischenstationen und auf Grund des Ölschocks war man 1975 bei 205 PS angelangt. Die Corvette war vom bärenstarken ,,muscle car" zu einem mit allen erdenklichen Extras ausrüstbaren Einkaufswagen sportlicher Hausfrauen geworden, wenn auch nur höchstens zwei volle Supermarkttüten unterzubringen waren. Kein Wunder, daß sich immer weniger Käufer für den übergewichtigen und im Vergleich zu früher lahmen Sportwagen interessierten. So sank die Verkaufszahl 1975 – es war das letzte Jahr des Stingray-Roadsters – auf magere 4629 Exemplare.
Es sollte nun fast zehn Jahre dauern, ehe eine völlig neue Corvette das Licht der Welt erblicken durfte, entwickelt von einem neuen Team von Konstrukteuren und Designern – unbelasted durch die Vergangenheit. Das betont niedrige und breite Coupé, das 1983 die sechste Generation dieses Sportwagens begründete, war wiederum von zahlreichen Show Cars inspiriert worden, und man muß zugeben, daß es ein sehr beeindruckender Wagen geworden ist. Er läßt die traditionellen Merkmale der Corvette, symbolisierte Kraft und sprungbereites Lauern, wieder aufleben. 1978 hatte man mit der intensiven Entwicklungsarbeit an diesem keilförmigen Wagen begonnen, der unter deutlichem Einfluß des Windkanals entstand. Er ist außer in der Breite in jeder Dimension kleiner als der Stingray, wirkt jedoch durch geschickte Verschiebung der Proportionen höchstens zierlicher. Endpunkt in der Styling-Entwicklung der neuen 230-PS-Corvette war schließlich 1986 die Präsentation der Roadster-Version, der ersten seit zehn Jahren. Es war dies ein modernes, gekonnt geformtes Fahrzeug – würdig, die Reihe beeindruckender offener Sportwagen von Chevrolet seit 1953 fortzusetzen.

## Bescheidener Anfang

Mit dem herkömmlichen Sechszylinder-Chevi-Motor – wenn auch mit 150 PS – war die Corvette kein allzu heißes Eisen. Und trotz des enormen Interesses, das der Wagen anläßlich seiner Präsentation geweckt hatte, gab es 1953–54 nur knapp 4000 Käufer für dieses Auto. Aber auch die Produktionsmöglichkeiten waren anfangs begrenzt.

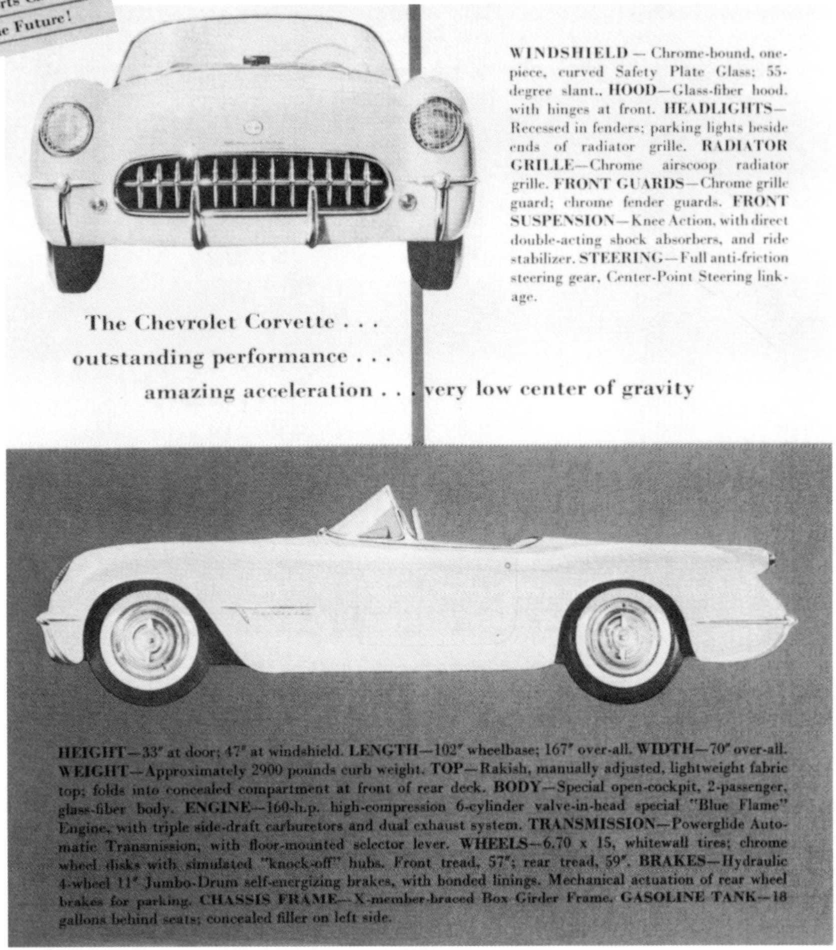


*Wenig aufwendig gibt sich der erste Prospekt, der anläßlich der Produktionsaufnahme im Sommer 1953 von General Motors herausgegeben wurde. Stolz wird auf die vielen Chromteile verwiesen.*

# Real driving comfort . . . the Corvette way!

SECURITY AND LUXURY for driver and passenger are the keynotes of the snug Corvette cockpit. Individual bucket seats have form-fitting foam rubber cushions. The floor is covered in soft carpeting, backed by sponge rubber. Large pockets and ash trays in doors also serve as arm rests. Beautifully balanced instrument panel includes key-turn starter, electric clock, tachometer, hooded radio speaker.

POTENT "BLUE-FLAME" 6 engine, with three side-draft carburetors, 8 to 1 compression ratio, and overhead valves, puts a flashing 155 horsepower under the throttle. It has a dual exhaust system, efficient cooling and lubrication, and a shielded electrical system . . . plus Chevrolet's traditional six-cylinder economy of operation and maintenance.

Im Modelljahr 1954 gab es die Corvette auch mit einem V8-Motor; ein Jahr später erhielt der Wagen Kurbelscheiben in den Türen und – auf Wunsch – ein Hardtop. Der Wagen begann für die Amerikaner interessant zu werden.

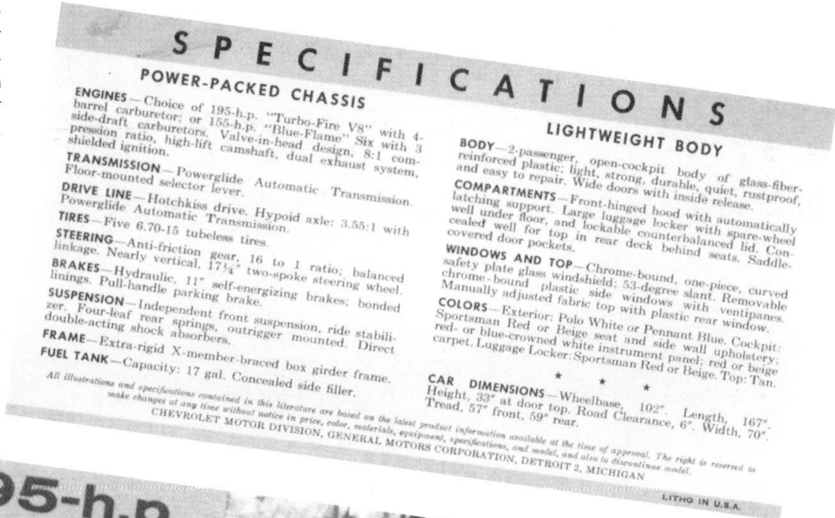

# SPECIFICATIONS

## POWER-PACKED CHASSIS

**ENGINES**—Choice of 195-h.p. "Turbo-Fire V8" with 4-barrel carburetor; or 155-h.p. "Blue-Flame" Six with 3 side-draft carburetors. Valve-in-head design, 8:1 compression ratio, high-lift camshaft, dual exhaust system, shielded ignition.

**TRANSMISSION**—Powerglide Automatic Transmission. Floor-mounted selector lever.

**DRIVE LINE**—Hotchkiss drive. Hypoid axle: 3.55:1 with Powerglide Automatic Transmission.

**TIRES**—Five 6.70-15 tubeless tires.

**STEERING**—Anti-friction gear, 16 to 1 ratio; balanced linkage. Nearly vertical, 17¼" two-spoke steering wheel.

**BRAKES**—Hydraulic, 11" self-energizing brakes; bonded linings. Pull-handle parking brake.

**SUSPENSION**—Independent front suspension, ride stabilizer. Four-leaf rear springs, outrigger mounted. Direct double-acting shock absorbers.

**FRAME**—Extra-rigid X-member-braced box girder frame.

**FUEL TANK**—Capacity: 17 gal. Concealed side filler.

## LIGHTWEIGHT BODY

**BODY**—2-passenger, open-cockpit body of glass-fiber-reinforced plastic; light, strong, durable, quiet, rustproof, and easy to repair. Wide doors with inside release.

**COMPARTMENTS**—Front-hinged hood with automatically latching support. Large luggage locker with spare-wheel well under floor, and lockable counterbalanced lid. Concealed well for top in rear deck behind seats. Saddle-covered door pockets.

**WINDOWS AND TOP**—Chrome-bound, one-piece, curved safety plate glass windshield; 53-degree slant. Removable chrome-bound plastic side windows with ventipanes. Manually adjusted fabric top with plastic rear window.

**COLORS**—Exterior: Polo White or Pennant Blue. Cockpit: Sportsman Red or Beige seat and side wall upholstery; red- or blue-crowned white instrument panel; red or beige carpet. Luggage Locker: Sportsman Red or Beige. Top: Tan.

* * *

**CAR DIMENSIONS**—Wheelbase, 102". Length, 167". Height, 33" at door top. Road Clearance, 6". Width, 70". Tread, 57" front, 59" rear.

*All illustrations and specifications contained in this literature are based on the latest product information available at the time of approval. The right is reserved to make changes at any time without notice in price, color, materials, equipment, specifications, and model, and also to discontinue model.*

CHEVROLET MOTOR DIVISION, GENERAL MOTORS CORPORATION, DETROIT 2, MICHIGAN

LITHO IN U.S.A.

*Links und rechts: Ein kleines Karton-Faltblatt von 1955, zweifarbig und ohne Übertreibungen. Als besonders sportlich galten Steinschlaggitter vor den Scheinwerfern.*

**A touch and she's up!**

The new power-operated fabric top (now available in white with black and beige, optionally keyed to exterior color) folds out of sight under its covered compartment. Wider windows increase visibility.

**Quick change! A roadster or coupe!**

A smartly-conceived plastic hardtop featuring a wrap-around rear window is available at extra cost. It is easily and securely installed with only five knurled set screws.

**Let it rain, let it snow**

Corvette offers the convenience of new "roll-up" regulators to quickly raise and lower the new side windows. Power window lifts, only one of Corvette's several new power features, are available at extra cost.

**New "going" look**

Now Corvette is even more adventurous-looking with graceful new fenderlines, new side panel and hood treatment and simulated knock-off type wheel covers.

**New "out-front" styling**

The raised forward portion of the fender houses an improved design headlight that projects forward to extend the fenderline and contributes to Corvette's rakish look.

**Eight-Jet take-offs!**

The Corvette's 265-cubic-inch V8 engine owes its greater horsepower to dual 4-barrel carburetion, higher compression ratio and new manifolds with "twin pipe" exhausts.

NEW

Corvette

Since its initial introduction in limited volum
Corvette has commanded the attention of spor
enthusiasts. Now its major new design factors v
welcomed as answers to the needs expressed b
well-informed group. Drivers who scorn the "

*Ähnlich aufgemacht wie der auf Seite 14/15 gezeigte Prospekt war dieses Faltblatt. Nur: Die Steinschlaggitter sind verschwunden, und die Flanken ziert eine kurvige Sicke. Die '56er Corvette ist attraktiv geworden . . .*

## SPECIFICATIONS

**ENGINE**

"Turbo-Fire Special V8." Valve-in-head design, 265-cubic-inch displacement, 3.75" bore x 3.0" stroke, 9.25:1 compression ratio. 225 horsepower at 5200 rpm. Torque 270 foot-pounds at 3600 rpm. Special high-lift camshaft, high-speed valve mechanism. Polished aluminum rocker covers. Dual four-barrel carburetion, buff aluminum racing-type air cleaners, special intake manifold. Full pressure lubrication system with full-flow oil filter.* High-power exhaust headers and full dual exhaust system. Shielded ignition, 12-volt electrical system. Engine precision balanced after assembly.

**TRANSMISSION**

Choice of special high-performance 3-speed close-ratio Synchro-Mesh (2.2:1 low and reverse, 1.31:1 second, 1.1 high) with high-capacity 10-inch coil-spring clutch, or optional Powerglide special automatic transmission.* Floor mounted gear or range selector.

**REAR AXLE**

High torque capacity axle; 3.55:1 ratio standard, 3.27:1 ratio optional with either Synchro-Mesh or Powerglide.

**CHASSIS**

Extra-rigid X-member-braced box girder frame. Independent coil front suspension with ride stabilizer. Self-lubricating four-leaf rear springs, outrigger mounted. Direct double-acting shock absorbers. Full anti-friction 16:1 ratio steering gear, balanced linkage. Competition-type steering wheel with three shock absorbing spring-steel spokes. Hydraulic 11-inch self-energizing brakes with new bonded linings, pull-handle parking brake. Suspended brake pedal. Choice of black or white sidewall* standard 6.70-15-4 ply tubeless tires or optional 6.70-15-4 ply high-speed nylon racing type.* Decorative wheel covers with simulated knock-off knobs. 17-gallon fuel tank with concealed side filler.

**BODY FEATURES**

Glass-fiber-reinforced plastic with sculptured side panels; light, strong, durable, quiet, rustproof, easy to repair. Distinctive embossed hood, front hinged, with automatic support, inside release. Simulated twin fender air scoops. Two-passenger compartment, large luggage locker with spare-wheel well under floor, concealed top well behind seats. Unique Corvette crossed-flag emblems on hood and trunk lid. Twin exhaust ports integral with rear bumpers. Chrome-bound, one-piece, curved safety plate glass windshield. Power-operated fabric top with wide plastic rear window. Richly trimmed quick-change hardtop* with rear-quarter windows and full-vision rear window.

**INTERIOR FEATURES**

Form-fitting vinyl-covered seats, individually adjustable, with safety belt.* Wide doors with built-in arm rest, push-button door handle, key lock, inside door release, swing-out door hinges. Choice of crank-operated or power* window lifts. Ash tray and glove compartment between seats; padded roll on instrument panel and doors, rubber-backed carpeting, metal door kick panels, sills, and step plates. Signal-seeking radio,* heater,* directional signals, electric clock, cigarette lighter, tachometer, outside and inside rear-view mirror, windshield washer.*

**COLORS**

Onyx Black with Red interior and Black or White top, Venetian Red with Red interior and Beige or White top, Cascade Green with Beige interior and Beige or White top, Aztec copper with Beige interior and Beige or White top, Arctic Blue with Beige or Red interior and Beige or White top, Polo White with Red interior and White or Black top.

**DIMENSIONS**

Wheelbase, 102". Length, 168.01". Overall height: top down, 49.20"; Convertible top, 51.09"; hardtop, 50.98". Height at door 32.55". Road clearance 6". Width, 70.46". Tread, 57" front, 59" rear.

*Optional at extra cost.

All illustrations and specifications contained in this literature are based on the latest information available at the time of publication approval. The right is reserved to make changes at any time without notice in price, color, materials, equipment and specifications.

**CHEVROLET MOTOR DIVISION • GENERAL MOTORS CORPORATION • DETROIT 2, MICHIGAN**

LITHO IN U.S.A.

*Die technischen Daten des Modells 1956. Die Vielfalt der Karosserie-Farben ist beachtlich!*

n convertible" and will accept nothing less than an hentic, competition-type sports car will quickly sense action is the keynote of the new Corvette V8. But venience is accented too, and you will find many ıry features seldom available in a car of this caliber.

*Live, Lithe, Luxurious Action*

*Ein Prospekt vom Sommer 1956, jetzt im Vierfarbdruck und mit einer Fülle interessanter Detailfotos. Vieles ist an der Corvette neu – siehe unten.*

## Close-up details reveal *New* Corvette Advancements

**A touch . . . and she's up!**

The new power-operated fabric top (now available in white with black and beige keyed to exterior color) folds out of sight under its plastic-lidded compartment behind the seat. Wider rear window and new-design side windows increase visibility.

**This . . . is for the "Box Boys"**

The new Corvette with a new floor-mounted "stick" shift close ratio Synchro-Mesh transmission. Here is the split-second up-shifting, down-shifting, close-ratio gear control demanded by the experts!

**Eight-Jet Carburetion for take-offs!**

The Corvette's 265-cubic-inch V8 engine owes its greater horsepower to dual 4-barrel carburetion, higher compression ratio and new manifolds with "twin pipe" exhausts.

**Avant-Garde styling with a touch of tradition**

The Corvette is a true sports car—not a scaled-down convertible. Now it's even more adventurous-looking with graceful new fender lines, new side panel and hood treatment and simulated knock-off type wheel covers.

**Quick change! A roadster or coupe!**

A plastic hardtop featuring a wraparound rear window is available as extra cost equipment. It is easily and securely installed with only five knurled set screws.

**Let it rain, let it snow!**

Corvette offers the convenience of roll-up regulators that quickly raise and lower the new windows. Power window lifts are available as an extra cost option.

**A hood full of "Horses" add a carload of Safety!**

The new "Turbo-fire V8" engine is a real life-saver when only sheer *passing power* can leave hazards behind and whisk you to safety. New cylinder heads up Corvette's compression ratio to 9.25 to 1!

**"Out-front" styling for looking ahead**

The raised-forward portion of the fender houses an improved design headlight that projects forward to extend the fender-line and contribute to Corvette's rakish look.

Das Titelblatt dieses Prospektes verheißt Tempo, doch eine Angabe zur Höchstgeschwindigkeit findet man nicht. Es genügte, von „high speed racing type tires“ und den vielen PS zu sprechen . . .

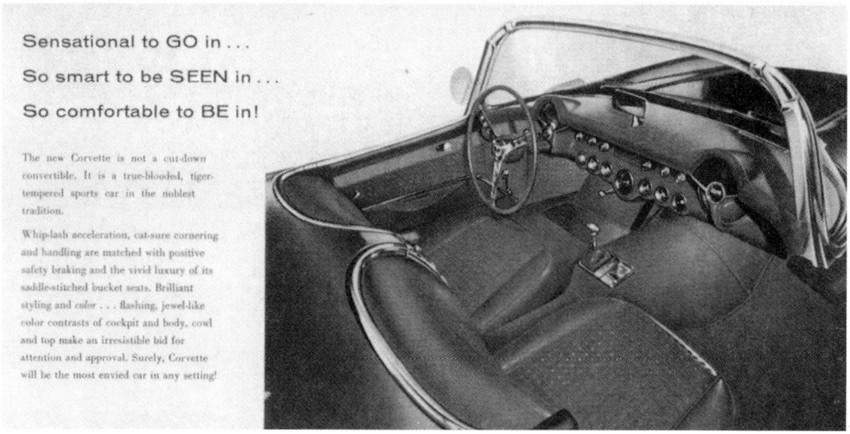

## Fuel Injection: Ein PS pro Kubikzoll

1957 präsentierte sich die Corvette mit Einspritzmotor. Es gab den V8 in zwei Varianten – mit 250 und mit 283 PS. Zugleich offerierte man ein Viergang-Getriebe mit Handschaltung. Eine neue Ära in der Geschichte des „einzigen amerikanischen Sportwagens" begann.

*Unten: Aus der Sammlung des Schülers Ralph Brown stammt dieser 1957er Prospekt. Er enthält erstmals Angaben zum neuen Einspritzmotor.*

*Rechts: Titelblatt eines 1959er Prospektes. Die Abbildungen werden flotter, die Umsatzzahlen besser.*

1959
Chevrolet
CORVETTE
AMERICA'S
SPORTS CAR

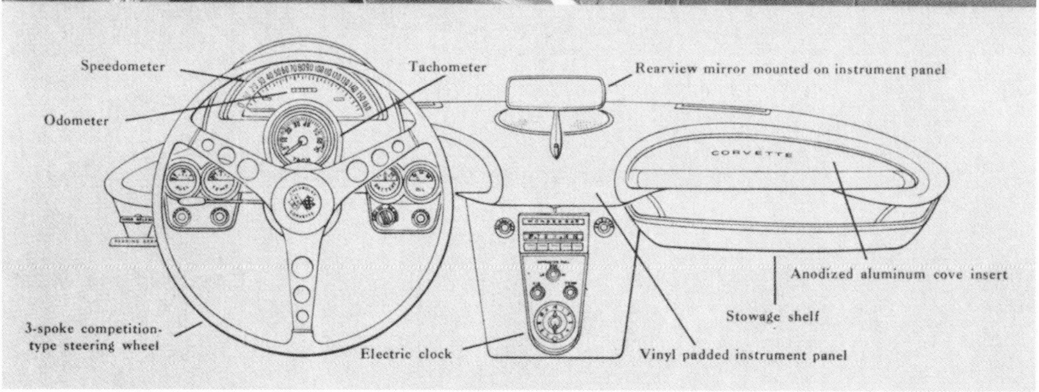

Speedometer
Tachometer
Rearview mirror mounted on instrument panel
Odometer
CORVETTE
Anodized aluminum cove insert
Stowage shelf
3-spoke competition-type steering wheel
Electric clock
Vinyl padded instrument panel

*Links Interieur 1959, rechts ein Prospekt von 1960.*

*Bilder einer Teststrecke enthält dieser Corvette-Prospekt von 1961. Beschleunigungs- oder andere „Speed"-Werte enthält auch er nicht. Worauf es dem Käufer ankam, läßt er dennoch deutlich erkennen. Die Gentlemen im Wagen müssen viel Spaß gehabt haben . . .*

1961

1962

*Oben: Der erste Sting Ray-Prospekt aus dem Jahre 1963. Motoren bis 360 PS sind lieferbar!*

*Links: Bemerkenswert aufgeräumt geht es nun im Cockpit der Corvette zu (1965).*

*Oben: Kurze Rückblende auf das Modelljahr 1962. Ein junges Traumpaar demonstriert auf bunten Seiten, wie schön das Leben an Bord der Corvette sein kann.*

Der Sting Ray (Stachelrochen) war wesentlich kantiger im Karosserie-Design als sein Vorgänger. Aber der neue Stil – eine Kreation Bill Mitchells – kam gut an. 10 919 Cabriolets und 10 594 Coupés wurden vom Modell-Jahrgang 1963 produziert. Später gab es allerdings eine deutliche Bevorzugung des offenen Wagens.

*Corvette Sting Ray Sport Coupe in Sebring Silver**

*Corvette Sting Ray Convertible in Riverside Red with Matching Hard Top*

*Corvette Sting Ray Convertible in Riverside Red with White Soft Top*

*Corvette Sting Ray Convertible in Riverside Red*

*Werbung für den Sting Ray 1963. Man bekommt den Wagen als Fastback-Coupé, als Cabrio mit Soft- oder mit Hardtop. Der Designer Arkus-Duntov kam mit dem neuen Styling den Vorstellungen vieler Amerikaner von einem modernen Sportwagen nahe.*

*Shown on Cover: Corvette Sting Ray Sport Coupe in Riverside Red*

*Corvette Sting Ray Convertible in Riverside Red*

# NEW GRACE AND ELEGANCE SILHOUETTE AMERICA'S SPORTS CAR

Corvette steps out smartly with an exhilarating new look for '63. A freshly elegant look that promises to lift the spirits of any buff who takes the wheel. It's the Corvette Sting Ray! Two sparkling new models, both pointing the way to a classic new concept in American sports car design.

Catch an eyeful of the Corvette newcomer, the exciting Sport Coupe. Graceful body panels and compound-curved side windows emphasize the miles-ahead aerodynamic design. Power-operated, retractable headlamps rotate out of sight to blend into the unobstructed hood line. Doors extend upward into the roof, adding armchair ease and comfort to sliding in and out. This is the Corvette Sting Ray Sport Coupe. Clean, taut, fresh in every detail.

You'll find the Convertible is jauntier than ever in '63. Uncluttered lines accented front and rear by trim wrap-around bumpers. The subtle contour of the smoothly molded rear deck. A gleaming aluminum grille, retractable head-lamps and a sweeping hood-length windsplit. Top up or down, graceful streamlining is the goal. And the Corvette Sting Ray Convertible achieves it.

# SURE-FOOTED CONFIDENCE ON THE TOUGHEST TRAILS

Two chassis innovations take much of the credit for the new Corvette's firm level ride, precise handling and maneuverability: a shorter 98" wheelbase and a unique rear suspension and newly designed rear axle. The rear suspension features a three-link independent system at each wheel. A radius arm running from the frame to the rear spindle support, a control rod attached to the differential and wheel spindle, and the double universal-jointed tubular axle shaft combine to provide nearly vertical wheel movement. One multi-leaf transverse spring (bolted to the differential carrier) extends from rear wheel to rear wheel and takes only vertical suspension loads. Braking, acceleration and lateral forces are transmitted by the radius arms, axles and control rods to the differential and the frame. The differential assembly is attached to the frame by a bolted-in crossmember; large rubber cups serve to dampen driveline vibrations. Small movements of the differential are taken up by universal joints on the one-piece propeller shaft. Corvette's new rear suspension in combination with improved coil-spring spherical joint front suspension produces better traction, a stable ride and excellent handling. You can add to tire and handling stability with Corvette's optional wide-rim cast aluminum wheels*. Vented-fin design for cooling; knock-off hubs for quick, simple wheel and tire changing.

For increased rigidity and protection, the Convertible has a steel framework surrounding the passenger compartment, while the Sport Coupe has overhead steel roof members as well.

*Optional Aluminum Wheels* with Knock-off Hubs.*

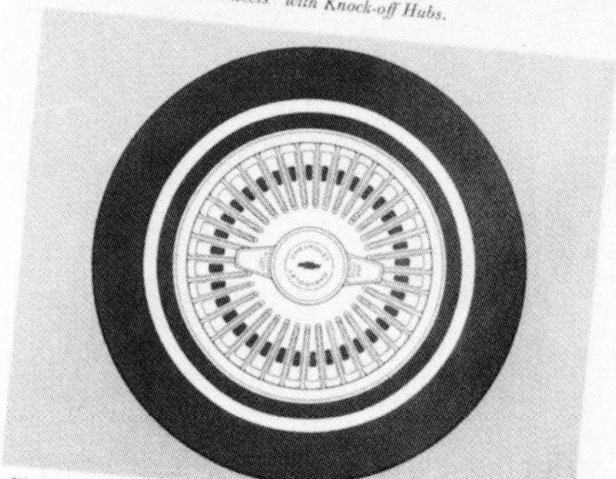

*Optional at extra cost.

*Leichtmetallfelgen mit Zentralverschluß – der letzte Schrei. Mit verkürztem Radstand und neuer Hinterachse verspricht G.M. ein völlig neues Fahrgefühl. Rechts: Erster 1965er Prospekt.*

1965
CORVETTE
STING RAY
CORVETTE

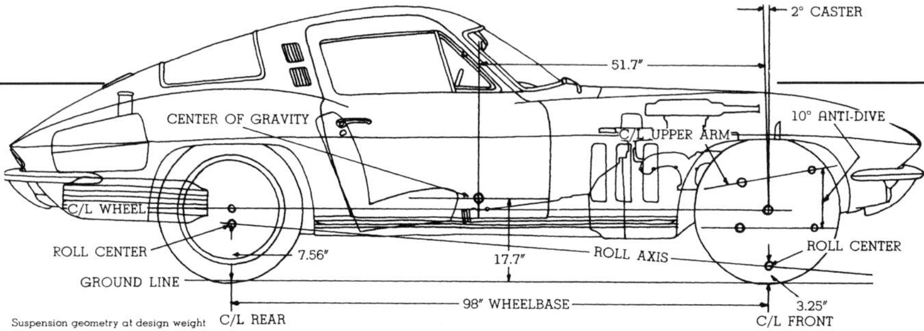


Suspension geometry at design weight

**DESIGN CONCEPT:** The key to Sting Ray's roadability and handling lies in its 4-wheel independent suspension and its greater rearward weight distribution. Chevrolet engineers chose the fully independent suspension so that optimum use could be made of the great power available. The rearward weight distribution makes it possible to achieve excellent handling and still maintain an acceptably gentle ride. The major vehicle masses — the heavier components of the Sting Ray—are located so that the suspension and steering systems can work with the design, rather than having to compensate for imbalance. The Sting Ray has been basically *right* from its original concept. Constant refinement and continual development have gone forward to make the 1965 Corvette Sting Ray more than ever one of the most deeply satisfying driving experiences available in the world.

**EXTRA-COST OPTIONAL EQUIPMENT**

The following equipment, shown previously in illustrations or described in the text, is available at extra cost for the Corvette Sting Ray. These options add driving pleasure, or prepare the car for special uses. They allow the Sting Ray owner to equip his car to his own particular tastes.

CHASSIS: Heavy-duty suspension. (Rear spring rate: 305 lb./in. Rear shock absorbers: 1⅜ in. Front spring rate: 550 lb./in. Front shock absorbers: recalibrated. Stabilizer bar: 0.94-inch diameter.) Positraction; power brakes; telescopic steering column; power steering; optional axle ratio; nylon tires; whitewall tires; cast aluminum wheels with wide-base six-inch rims and knock-off hubs.

ENGINE: Off-road service exhaust system; exposed, side-mounted exhaust system; 36-gallon fuel tank (Sport Coupe only). (Note: when 36-gallon fuel tank is fitted, luggage compartment is partially carpeted); transistor ignition and voltage regulator; 300-horsepower engine; 350-horsepower engine; 365-horsepower engine; 375-horsepower Ramjet Fuel Injection engine.

TRANSMISSION: 4-Speed; 4-Speed close ratio; Powerglide.

BODY: Soft-Ray tinted glass; backup lights and non-glare inside rearview mirror; Four-Season air conditioning; genuine leather seat upholstery; wood-rimmed steering wheel; electric windows; AM/FM pushbutton radio with remote control power antenna; removeable hard top. (Convertible only. You can, however, specify either hard top or soft top at no extra cost—or order both with hard top at extra cost.)

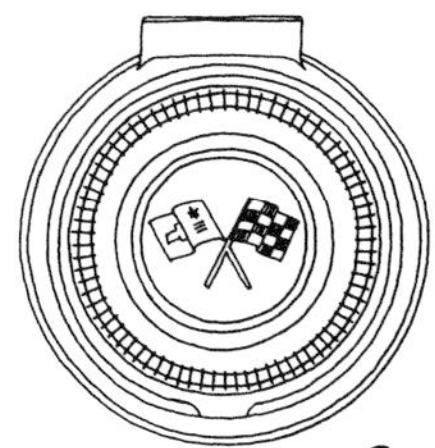

Gas filler door emblem

**from rubber to roof, a sports car...**

**for performance -and style-minded individuals.**

*Bilder aus einem 1965er Prospekt. Er enthält ausgezeichnete Fotos und detaillierte Zeichnungen.*

CORVETTE

CORVETTE

Die Karosserie der '65er Corvette besteht laut Katalog aus 31 Kunststoff-Teilen. Rechts: Abbildung aus einem '66er Prospekt.

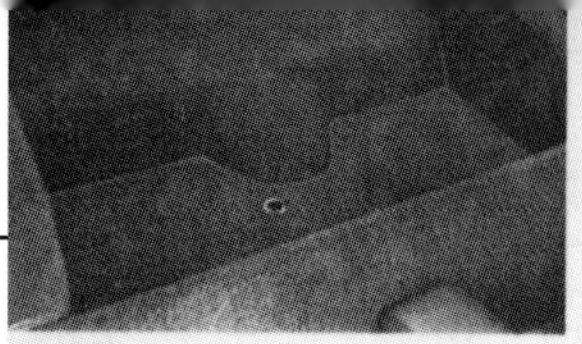

The sport coupe's fully carpeted luggage area

Right: Enthusiast's delight: 4-Speed box

Seat belts with retractors are standard

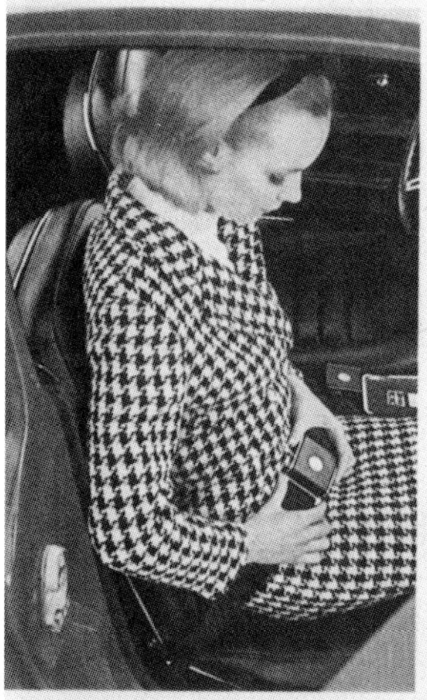

Pushbutton AM/FM all-transistor radio

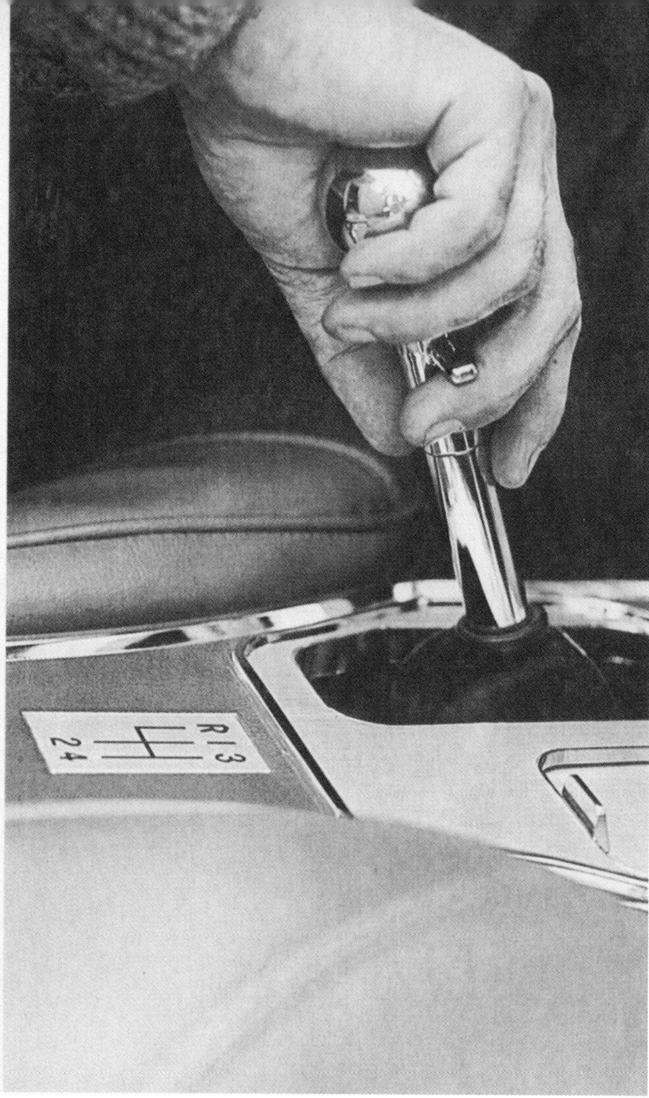

*Als sportlich galt die Handschaltung, aber man erhielt auf Wunsch auch eine Automatik. Mit Zweipunkt-Gurten gab man sich indes noch immer zufrieden.*

CORVETTE

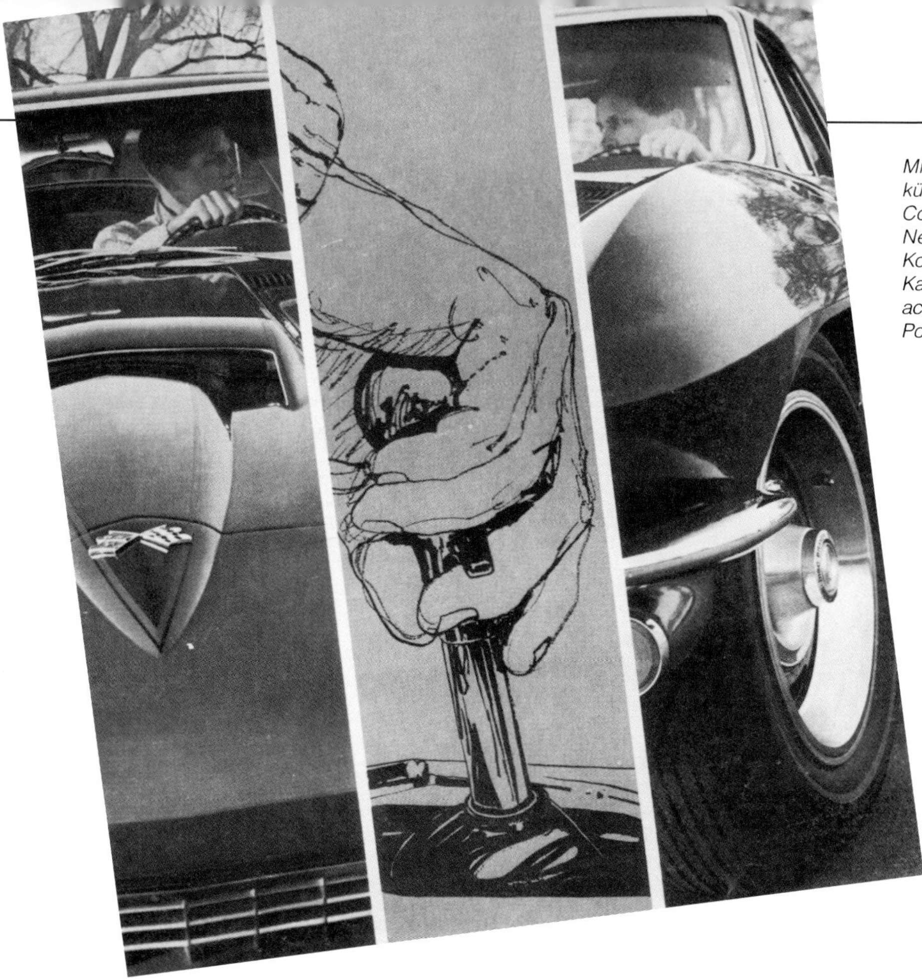

*Mit flotten Zeichnungen kündigt dieser '67er Corvette-Prospekt die Neuerungen an. Eine Kombination von zehn Karosseriefarben und acht Varianten der Polsterung wird offeriert.*

**'67 Corvette**
**BY CHEVROLET**

# energetic

By definition, energy is "internal power with the capacity of acting." You can define the kind of energetic action you want in your Corvette with the help of an expanded lineup of V8s for '67, backed by three transmissions and five rear axle ratios.

The big attraction for Corvette fanciers is the change that's come over the 427 Turbo-Jets. Now there are three of these top-performing 427-cu.-in. engines under a special bubble hood – two of them brand-new and equipped with three 2-barrel carburetors in a row. Highest rated member of this trio is the 435-hp version which features mechanical valve lifters and a special-performance camshaft. It has a compression ratio of 11.0:1.

The 400-hp 427 also features triple 2-barrel carburetion but uses hydraulic lifters and a high-performance camshaft for quiet operation. Finally, there's a 390-hp 427 Turbo-Jet with 4-barrel carb, hydraulic lifters and high-performance cam. These two 427s have a 10.25:1 compression ratio.

Special features make the three 427 Turbo-Jet engines extremely durable and efficient. To name a few: extra-wide-base main bearing caps that add to the support of each of the five main bearings; premium aluminum main and connecting rod bearings plus construction which produces extended durability; big valves that help cylinder heads breathe quickly and freely; individual intake and exhaust ports plus other refinements all help to increase the volume of combustibles and improve exhausting.

All 427 V8s are bored to 4.251 inches and travel through a 3.76-inch stroke. Combustion chambers are of modified-wedge design. Resulting forces of combustion react more to push the piston down rather than heat the sidewalls of the chamber.

Standard engine in a '67 Corvette is the 300-hp 327-cu.-in. V8. And if you need more, you can order the 350-hp version of the same 327 block. The 300-hp engine utilizes a general-performance cam and has a compression ratio of 10.0:1; the heftier 327 V8 has a high-performance camshaft, 4-barrel carburetor with special aluminum intake manifold and an 11.0:1 compression ratio. Both have hydraulic valve lifters.

Oil capacity for the standard Corvette engine is five quarts – filter included. For the extra-cost engines, six quarts of oil are required with filter.

For the '67 Corvette, a 3-Speed fully synchronized gearbox is coupled with the 300-hp 327-cu.-in. engine as standard equipment. A 4-Speed with 2.52:1 ratio 1st gear may be ordered with 300-, 350-, 390-, or 400-hp engines; a 2.20:1 ratio 1st gear 4-Speed may be specified with 350-, 390-, 400-, and 435-hp V8s. This year, for more leisure-minded sports car buffs, Powerglide automatic transmission can be ordered with the 300-, 390-, and 400-hp engines.

Five rear axle ratios are available, depending on the engine you select, to provide you with smooth going or rugged performance. A Positraction rear axle, which can be specified with any Corvette ratio, routes engine power to the rear wheel with the greatest traction. This is especially valuable on muddy, slippery or other irregular surfaces, in sand and snow. Check the axle ratio chart at right to determine rate of speed with various axle ratios.

**1967 STING RAY POWER TEAMS**

| Engine Bore & Stroke | Horsepower & Torque at RPM | Carburetion & Induction System | Comp. Ratio | Cam & Lifters | Transmissions | Axle Ratios Standard | Axle Ratios Positraction |
|---|---|---|---|---|---|---|---|
| **STANDARD ENGINE** | | | | | | | |
| 327-cu.-in. V8 | 300 @ 5000 | 4-Barrel | 10.0:1 | General Performance | 3-Speed (2.54:1 Low) | 3.36:1 | 3.08:1<br>3.36:1 |
| 4.00 x 3.25 ins. | 360 @ 3400 | High-Flow Air Cleaner | | Hydraulic | 4-Speed (2.52:1 Low) | 3.36:1 | 3.08:1<br>3.36:1 |
| | | | | | Powerglide | 3.36:1 | 3.36:1 |
| **EXTRA-COST OPTIONAL ENGINES** | | | | | | | |
| 327-cu.-in V8 | 350 @ 5800 | 4-Barrel | 11.0:1 | High Performance | 4-Speed (2.52:1 Low) | 3.36:1 | 3.36:1<br>3.55:1 |
| 4.00 x 3.25 ins. | 360 @ 3600 | High-Flow Air Cleaner | | Hydraulic | 4-Speed (2.20:1 Low) | 3.70:1 | 3.70:1<br>4.11:1 |
| 427-cu.-in. V8 | 390 @ 5400 | 4-Barrel | 10.25:1 | High Performance | 4-Speed (2.52:1 Low) | 3.08:1* | 3.36:1 |
| 4.251 x 3.76 ins. | 460 @ 3600 | High-Flow Air Cleaner | | Hydraulic | 4-Speed (2.20:1 Low) | 3.36:1* | 3.08:1<br>3.55:1 |
| | | | | | Powerglide | 3.36:1* | 3.70:1 |
| 427-cu.-in. V8 | 400 @ 5400 | Triple 2-Barrel | 10.25:1 | High Performance | 4-Speed (2.52:1 Low) | 3.08:1* | 3.36:1 |
| 4.251 x 3.76 ins. | 400 @ 3600 | High-Flow Air Cleaner | | Hydraulic | 4-Speed (2.20:1 Low) | 3.36:1* | 3.08:1<br>3.55:1<br>3.70:1 |
| | | | | | Powerglide | | |
| 427-cu.-in. V8 | 435 @ 5800 | Triple 2-Barrel | 11.0:1 | Special Performance | 4-Speed (2.20:1 Low) | 3.55:1* | 3.36:1*<br>3.70:1*<br>4.11:1* |
| 4.251 x 3.76 ins. | 460 @ 4000 | High-Flow Air Cleaner | | Mechanical | | | |

*Available as Positraction only

**MILES PER HOUR PER 1000 ENGINE RPM IN FINAL DRIVE**

(Figures are estimated without allowance for tire expansion or slippage.)

3.08:1 = 25.6  3.36:1 = 23.1
3.55:1 = 22.2  3.70:1 = 21.3
4.11:1 = 19.2

Try a '67 Corvette as soon as you can. You'll find there's no energy wasted.

EXTRA-COST OPTIONS AND CUSTOM FEATURES* TO MAKE DRIVING A CORVETTE EVEN MORE PLEASURABLE

350-hp Turbo-Fire 327; 390-, 400- or 435-hp Turbo-Jet 427. An automatic Powerglide or 4-Speed fully synchronized transmission. Four-Season air conditioning. Power brakes, power steering, power windows. AM/FM pushbutton radio. Removable hardtop for convertible. Black vinyl roof cover for removable hardtop. Positraction rear axle. Leather seat trim. Shoulder belts. Whitewall or red-stripe tires. Soft-Ray tinted windows and/or windshield. Full-transistor ignition system (not available with standard engine). Heavy-duty brakes. Special-performance front and rear suspension. Off-road exhaust system (not available with Powerglide). Dual side-mounted off-road exhausts. Telescopic steering wheel. Emergency road kit. Compass. Gas tank filler cap lock. Removable floor mats. 36.5-gallon gas tank for sport coupe only. Convertible deck lid luggage carrier plus strap. Convertible deck lid ski rack. Strato-ease headrests. Speed warning unit. Portable hand spotlight. Special cast-aluminum wheels.

*Check your Chevrolet dealer for these and other extra-cost items along with their model application and availability with other equipment.

9

*Motorvarianten, die keine Wünsche offen ließen, und dementsprechende Getriebe- und Hinterachs-Abstufungen. Die Aufpreise bewegten sich für derlei Extras in zivilen Bereichen . . .*

*Links: Titelseite des ersten Prospekts für das Modelljahr 1968. Der innen wie außen gänzlich überarbeitete Wagen wirkt extrem bullig, wenn nicht aggressiv. Dreipunktgurte sind im Coupé jetzt Standard.*

*Hide-A-Way windshield wiper-washer system closed . . . and open.*

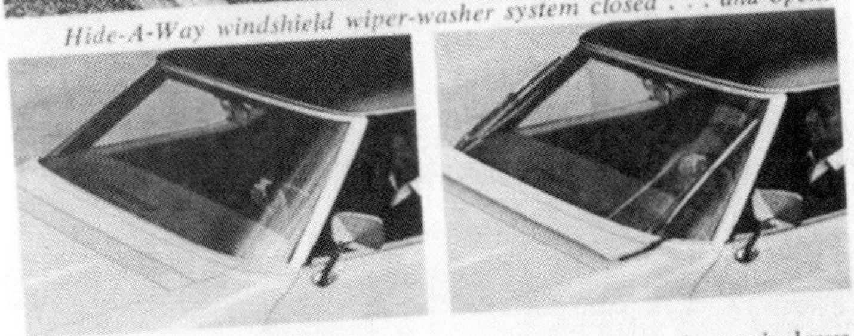

For comfort, air flows through Corvette with full door windows closed, thanks to the new Astro Ventilation system with vent-ports in the driver and passenger sides of the instrument panel. For protection, there are safety features like never before (some are listed on page 11). For long-lasting good looks, choose from ten Magic-Mirror finishes. *Corvette '68 . . . all different all over.*

*Rechts: Neu sind auch viele Details wie die versenkten Scheibenwischer oder das Belüftungssystem. Die Presse ermittelte den Wagen als „The Best Allround Car" des Jahres – zum viertenmal. Die Abbildungen auf den nachfolgenden Seiten entstammen dem gleichen Prospekt.*

*Corvette Sting Ray Coupe interior with high-back slim-tapered bucket*

# Settle down!

The Corvette cockpit is made for the traveling duo. Thin tapered and contoured buckets are very comfortably high-backed. Supple all-vinyl upholstery is standard, or you can order genuine leather. Between the seats the center console houses the parking brake lever, gear shift, cigarette lighter and ashtray, thumb-wheel heater controls, air vent controls, and light monitoring system indicators (to check operation of important outside running lights from inside through space-age fiber optics). The console also stows seat belt buckle straps—twin sets with pushbutton buckles are standard. Coupe equipment also includes twin shoulder belts.

Riding just above the console is the recessed instrument cluster. Here's where the engine tale is told. Water temperature, oil pressure and fuel gauges along with an ammeter and rally clock are positioned for easy reading. Windshield wiper-washer controls are also in console. And if you order an AM/FM radio—available or without FM Stereo multiplex—controls are mounted here zontally. Joining in are new features like door ajar flashe seat belt reminder light. Directly in front of the drive the speedometer, tachometer, trip odometer, light controls, tion lock, high beam indicator, turn signal indicator, brake s warning light—and a 3-spoke steering wheel that looks like

Luxurious deep-twist carpeting looks and feels great. It

*Shown on cover: Corvette Sting Ray Coupe*

*Corvette Sting Ray Coupe with roof sections in plac*

# Go hardtop!

The Corvette Sting Ray is for those who appreciate the true sports kind of car – and even for those who don't right now. It's that special kind of an automobile that comes along about once every generation to totally arrest the imagination of car buffs. In 15 years of Corvettes the car has not only driven into prominence in the sports car field, but has also been a forerunner of some exciting and practical automobile innovation. For 1968, the mechanics of Sting Ray have been improved and refined (it's still basically made for people who feel that the best part of living is driving), and this, obviously, is a most inspiring year for design.

Consider: the '68 Sting Ray Coupe is a hardtop and more. Uncommon removable sections over pilot and navigator lift out for open air moving. A nearly vertical glass rear window tucks out of the way into a neat compartment in the luggage area. The effect is a flow-through cochère roof that's never been seen on an American production sports car until now.

Long, low profile with blunt styling brings up the rear per th continental GT tradition. The aerodynamic design features spoiler back there, too. Behind front wheels, functional louve help keep the horses cool. Wraparound front and rear bumpe plus line-smoothing hideaway features help make Corvette trim one style-wise. On the nose end, vacuum-operated hea lights glide open automatically when lights are turned on. Win shield wipers aren't around when they shouldn't be. They' hidden under a power-operated panel which actuates whe

1969 CORVETTE
Putting you first, keeps us first
CHEVROLET

*Titelblatt des Ankündigungsprospektes für 1969 und Interieuraufnahme. Studiotricks und brillante Close-ups wechseln sich ab (nachfolgende Seiten).*

grabbing 15-inch wide ovals plus new 8-inch wide wheels and wider tread give Stingray a better grip . . . and an even bolder appearance. Other subtle but significant refinements include a stronger, more rigid frame for improved ride and handling, new maximum security 3-way locking system on the steering column and unique headlight washers that even clean in motion with jets of water.

Shown here are the ways you can get it: Coupe with removable hardtop sections; Convertible with soft top or removable hardtop. *For a more complete list of Corvette extra-cost Options and Custom Features illustrated or described throughout this book, see Page 11.*

Convertible with top down.

Vinyl-covered removable hardtop.

Headlight washers.

5

V8's in addition to the standard 300-hp V8. Prefer 4-Speed? You have a choice of gear ratios offered with all engines except the standard 300-hp and the 435-hp V8.

New Corvette power team features for '69 include new, larger 350 CID 300-hp engine as standard, or a 350-hp version is available, sturdier engine construction on the 350 CID, thicker bulkheads and main bearing caps, firmer crankshaft clamping with longer washer head bolts, extra-firm 4-bolt main bearing cap attachment and improved cooling system.

We even have a special engine (L-88) that we don't recommend for street use.

Transmissions: Standard 3-Speed fully synchronized; 4-Speed fully synchronized (wide-range or close-ratio); Turbo Hydra-Matic which operates automatically unless you want to shift it—through three forward ranges.

Of course, four-wheel disc brakes are still standard on Corvette—and in stopping ability, fade resistance and modulation they're very much appreciated by just about anybody interested in sports cars. Other features you get with Corvette include a new anti-interference ignition system, a high-output Delcotron generator, high-capacity energizer type battery, rustproof fiber glass body—and everything put together with a thoroughness and attention to detail that Corvette owners fully understand.

Standard equipment safety and security features on the 1969 Corvette include: energy-absorbing steering column □ seat belts with pushbutton buckles □ shoulder belts with inertia retractors (Coupe only) □ head restraints □ passenger-guard door locks □ safety door latches and hinges □ four-way hazard warning flasher □ dual master cylinder brake system with warning light and corrosion-resistant brake lines □ tire safety rims □ folding seat back latches □ dual-speed windshield wipers and washer □ dual-action safety hood latch □ outside rearview mirror □ back-up lights □ side marker lights and parking lights that illuminate with headlights □ energy-absorbing instrument panel □ padded sun visors □ reduced-glare instrument panel top, inside windshield mouldings, steering wheel hub and windshield wiper arms and blades □ wide inside day-night mirror with deflecting base □ lane-change feature in direction signal control □ safety armrests □ thick-laminate windshield □ soft, low-profile window control knobs □ smooth contoured door and window regulator handles □ anti-theft ignition key warning buzzer □ starter safety switch (to prevent engine starting when in gear) on all transmissions □ snag-resistant steering wheel hardware □ non-projecting wheel trim □ improved fuel tank retention □ headlight aiming access provision □ fail-safe concealed headlights □ anti-theft ignition, steering and transmission lock.

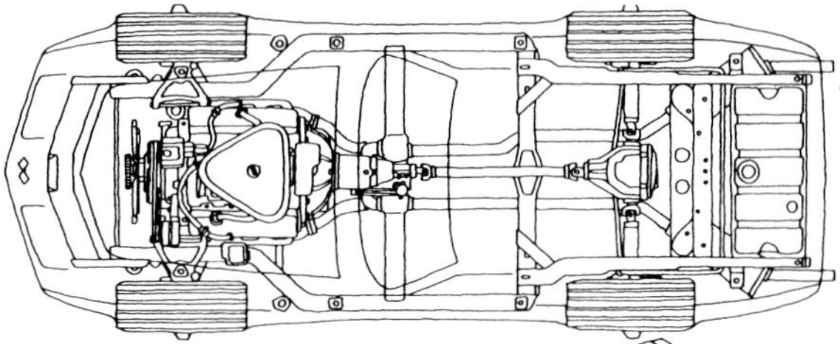

Top view of chassis.

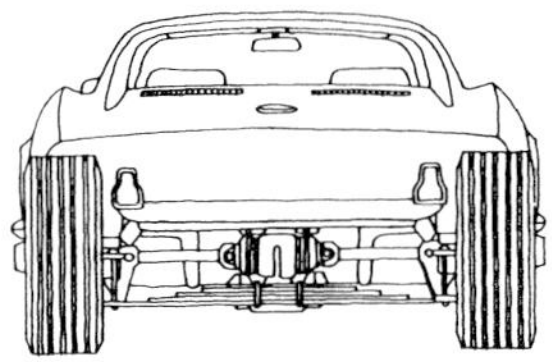

Fully independent rear suspension.

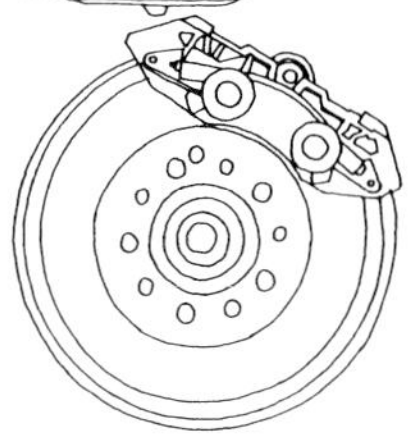

Advanced design disc brakes.

| Engine | Transmission | Rear Axle Ratio (:1) | | | | | | | |
|---|---|---|---|---|---|---|---|---|---|
| | | Without Air Cond. | | | | With Air Cond. | | | |
| | | Std | Optional | | | Std | Optional | | |
| | | | Econ | Perf | Spcl | | Econ | Perf | Spcl |
| BASE ENGINE | | | | | | | | | |
| 300 HP Turbo-Fire 350 350-Cu.-In. V8 | 3-Speed (2.54:1 Low) | 3.36 * | 3.08 ** | | | 3.36 * | 3.08 ** | | |
| | 4-Speed (2.52:1 Low) | | | | | | | | |
| | Turbo Hydra-Matic | 3.08 ** | | | | 3.08 ** | | | |
| OPTIONAL ENGINES | | | | | | | | | |
| 350 HP Turbo-Fire 350 350-Cu.-In. V8 RPO L46 | 4-Speed (2.52:1 Low) | 3.36 * | | 3.55 ** | | 3.36 * | | 3.55 ** | |
| | 4-Speed (2.20:1 Low) | 3.70 * | | 4.11 ** | | 3.70 * | | 4.11 ** | |
| 390 HP Turbo-Jet 427 427-Cu.-In. V8 RPO L36 | 4-Speed (2.52:1 Low) | 3.08 ** | | 3.36 ** | | 3.08 ** | | | |
| | 4-Speed (2.20:1 Low) | 3.36 ** | 3.08 ** | 3.55 ** | 3.70 ** | Air Conditioning Not Available | | | |
| | Turbo Hydra-Matic | 3.08 ** | 2.73 ** | | | 3.08 ** | 2.73 ** | | |
| 400 HP Turbo-Jet 427 427-Cu.-In. V8 RPO L68 | 4-Speed (2.52:1 Low) | 3.08 ** | | 3.36 ** | | 3.08 ** | | | |
| | 4-Speed (2.20:1 Low) | 3.36 ** | 3.08 ** | 3.55 ** | 3.70 ** | Air Conditioning Not Available | | | |
| | Turbo Hydra-Matic | 3.08 ** | 2.73 ** | | | 3.08 ** | 2.73 ** | | |
| 435 HP Turbo-Jet 427 427-Cu.-In. V8 RPO L71 | 4-Speed (2.20:1 Low) | 3.55 ** | 3.36 ** | 3.70 ** | 4.11 ** | Air Conditioning Not Available | | | |
| | Turbo Hydra-Matic | 3.08 ** | 2.73 ** | 3.36 ** | | | | | |
| SPECIAL HIGH PERFORMANCE ENGINE (OFF-ROAD APPLICATION ONLY) | | | | | | | | | |
| 430 HP Turbo-Jet 427 427-Cu.-In. V8 RPO L88 | H.D. 4-Speed (2.20:1 Low) | 3.36 ** | 3.08 ** | 3.55 ** | 3.70 4.11 4.56 ** | Air Conditioning Not Available | | | |
| | Turbo Hydra-Matic | 3.08 ** | 2.73 ** | 3.36 ** | | | | | |

*Positraction available. **Positraction required.

Corvette Stingray Coupe
CHEVROLET
Litho in U.S.A.—D-67141
GM

*Links: 1969er Werbung – mit viel Black Magic. Rechts: 1970 hieß es, man erhalte mehr als nur ein Auto, wenn man eine Corvette kaufe. Ob die Dame im Hintergrund gemeint war?*

## Mission Control Center '70.

Cape Kennedy hasn't got anything on us. Corvette's cockpit is as heavy on instrumentation as it is on comfort.

First, you sit yourself down in one of the high-backed contoured bucket seats featuring our new integrated head restraints. The shoulder belts (standard on coupe) are guided through a slot in each seatback for neat, secure positioning.

You rest your feet on soft carpeting that not only stretches wall to wall but over the entire rear stowage area with its hidden compartments for valuables, battery and tools.

You check out the central command console before your countdown. Tach, ammeter, water temp, oil pressure, fuel gauge, brake warning light, running light monitors—you name it, it's there.

Once under way, you can enjoy Astro Ventilation, which routes outside air in, through your choice of high or low vents. A rear deck vent exhausts stale air, for a constant flow, even with the windows closed.

New for '70 is a Custom Interior you can order with black or saddle leather seats, plush cut-pile carpeting and the rich look of wood on door panels and console.

The standard all-vinyl interior comes in black, saddle, dark brown, dark green, blue and red.

No matter if your mission's only a trip to the drive-in, the interior of the 1970 Corvette is designed to put you in complete control.

Chevrolet Corvette Custom Interior and Tilt-Telescopic steering column.

9

*Links: Innenseite aus dem 1970er Corvette-Prospekt. Rechts: Starke Nummer – 425 SAE-PS werden geboten. Vorsorglich wird der Wagen auf einer (leeren) Rennstrecke gezeigt.*

## Strength in numbers for 1971 - 270, 330, 365, 425 hp.

Every Corvette is built tough right from the start. Our basic V8, the 270-hp Turbo-Fire 350, features an 8.5:1 compression ratio, a 4-barrel carb, cast aluminum alloy pistons and a precision-formed crank. With this, and all other Corvette engines, you get Corvette's dual exhausts with rectangular outlets and Positraction differential standard.

Go one step up to the 330-hp Turbo-Fire 350, and you get a 9.0:1 compression ratio, impact-extruded aluminum alloy pistons, finned aluminum rocker covers, a special-performance cam, mechanical lifters and a forged steel-alloy crank. Plus an aluminum intake manifold, high-domed hood, transistor ignition.

Next step up. Our 365-hp 454-cu.-in. Turbo-Jet V8. This engine with 8.5:1 compression ratio uses a dual-snorkel air cleaner with a chromed cover and an extra-large oil pan. You also get a high-domed hood, a larger diameter front stabilizer bar, heavier duty rear wheel spindle support arms, a rear suspension stabilizer bar, a larger capacity radiator, dual pulleys for the fan and water pump and a higher performance starter motor.

Also available: our 425-hp Turbo-Jet 454 with a large 4-barrel carb, high-performance cam, mechanical lifters, 9.0:1 compression ratio and aluminum cylinder heads. Plus chassis items included with the 365-hp V8.

Transmissions: standard gearbox is a floor-mounted, wide-range 4-Speed, 2.52:1 low gear. Also available, a close-ratio version with a 2.20:1 low (close ratio not available with 270-hp engine, wide range not available with 425-hp engine). A Special 4-Speed close ratio box is available with the 425-hp V8.

With the 270-, 365-, and 425-hp engines, you can order Turbo Hydra-matic.

Chevrolet Corvette Stingray Convertible with removable hardtop

*Aufnahmen aus einem '73er Katalog. Interessant, daß man „Stingray" jetzt nicht mehr in zwei Worten schreibt.*

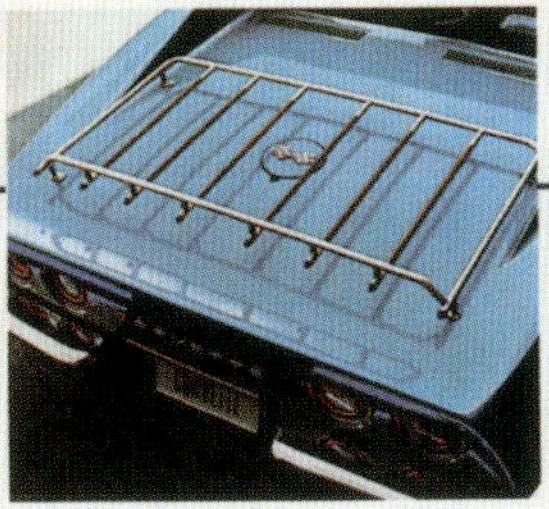

**orvette Stingray Convertible**
s being under the weather a
ure. All you have to do with the
g top is unlatch it, raise it, drop it,
t and forget it. The deck cover
s it all. Top available in both
and white.

an also order a removable hard-
s well worth having. For winter,
riety and for saving your
g top for a sunny day.

**orvette Instrument Panel** is
ike a cockpit than a dashboard.
right: a big speedometer and an
y big tachometer. The five other
s tell you the coolant temperature,
ssure, amperage, fuel level and
e. Above the gauges on the con-
ontrols for the windshield washers
ipers. Below, heating and venti-
controls. Within easy reach is the
M radio or AM/FM/Stereo radio
an order. With one speaker for
nd one for your co-pilot.

**The Standard Corvette Interior** is up
to pretty high standards.
The high-backed contoured bucket
seats are virtually handmade. They're
cushioned with thick foam and have
built-in head restraints. Fold either sea
forward and you're into Corvette's
fully carpeted, illuminated luggage
area. Two storage compartments are
concealed behind the seats. A third
compartment conceals the battery. Th
remainder of the passenger area is
trimmed in rich padded vinyl with dee
twist carpeting on the floor. Available
in black, blue, medium saddle, dark
saddle and red.

**The Available Custom Corvette
Interior** (shown at left). The seat pane
are genuine leather with matching
vinyl trim. There are wood-grained
vinyl accents on the doors and console
Underfoot, deep cut-pile nylon car-
peting. You're literally riding in the la
of luxury. Black, medium saddle or
dark saddle.

BUILDING A BETTER WAY TO SEE THE U.S.A.
Chevrolet

# OPTIONS

*Corvette-Werbung 1974 – einmal ganz anders. Fotos und leichte Aquarelle geben einen ungewöhnlichen Kontrast. In kleingedruckten Zeilen wird darauf hingewiesen, daß man auch Ledersitze haben kann.*

## FOR THE BODY

**Removable Hardtop** (for convertible)—The way to give your convertible top a vacation during cold weather. And yourself the added warmth and quiet of a hardtop. It's styled to enhance the lines of the Corvette and can be ordered to match the exterior color, or with an available black vinyl roof cover for a more formal look.

**Luggage Rack** (dealer-installed)—It's handsome enough to be strictly a dress-up item. But isn't it nice that it also lets you store the occasional overflow from your luggage compartment?

**Also Available** (from your dealer)— • Right-hand outside mirror • Non-vented locking gas cap.

## FOR THE CHASSIS

**Turbo-Fire 350-4 Special V8**—When we say that we've given this 350 a special carb, cam, crank, cylinder heads, pistons, rods, even finned aluminum rocker arm covers, we still haven't said it all. Designed to come on in the higher RPM numbers, the Special is obviously an engine for the enthusiast.

**Turbo-Jet 454-4 V8**—Like they say: "Nothing beats inches." And with 454 cubic inches of displacement, the Turbo-Jet provides a lusty torque output over the entire RPM operating range. This engine, like all 1974 Corvette engines, features a new multi-louvered radiator for improved cooling in city traffic conditions.

**Console-Shift Turbo Hydra-matic Transmission**—Think of this automatic three-speed as a box with a dual personality. Leave it in "Drive" range when you've got other things on your mind. Shift it manually with the console T-handle when special acceleration or engine braking needs are called for. Available with all engines.

**Close-Ratio 4-Speed Manual Transmission**—Fully synchronized with a 2.20:1 first gear ratio, this gearbox provides

*Schwarz und stark – und jetzt mit dem beliebten T-Top!*

# The essence.

This is what makes your Corvette what it is: America's only true production sports car.

**ON THE OUTSIDE.**

Corvette's low-slung aerodynamic styling and fiberglass body need no introduction. Also:

• Lightweight, remove-and-store roof panels convert your Corvette in seconds from closed- to open-top coupe, or vice versa. This year we've made provisions to stow the panels on the available deck-lid luggage carrier if you wish, to give more interior luggage room (see back cover).
• Energy-absorbing bumpers covered by soft, molded urethane in body color.
• Dual molded bumper guards, front and rear.
• Flush, retracting dual headlamps; dual-unit circular taillights.
• Soft-Ray tinted glass.
• Black windshield posts give new "thin pillar" look.
• 15" x 8" steel wheels with bright hub covers and trim rings.
• Hide-A-Way windshield wipers.
• New exterior colors: yellow, black, dark red, orange, dark blue, light blue metallic, tan buckskin. Continued: classic white, silver and medium red.

**THE COCKPIT.**

It's intimate as only a two-seater can be, yet functional as a sports car must be.

• Two-level instrumentation. Tachometer and speedometer with trip odometer are centered in front of the driver's eyes, high on the panel (just below the driver's road view) on two large circular dials. Voltmeter, coolant temperature, oil pressure, fuel level and electric clock are on a separate panel, in smaller, aircraft-styled dials, at the forward end of the console.
• Center console puts the following controls at your right hand: shift selector (with new leather boot); heating/ventilation plus ashtray and lighter. Controls for available air conditioning, power windows, Electro-Clear rear window defogger with warning light are also located on the console. New features include a handy coin tray and a soft vinyl housing over the parking brake handle.
• Four-spoke, padded-vinyl-covered steering wheel is now positioned two inches farther from driver to ease entry and exit.
• New column-mounted "Smart Switch" puts these controls at your left hand: turn signals, hi-lo headlight beam selection, windshield wipers and washers.
• Padded sun shades now swivel to provide some glare protection from side as well as front.
• Overhead and underpanel courtesy lights actuate automatically when either door opens, or manually with headlight switch.
• Panel-mounted map pockets.
• Extensive sound insulation.
• Nylon, cut-pile, color-keyed carpeting for passenger and luggage compartment.
• Luggage compartment behind seats and two storage chambers (one lockable for valuables), featuring increased ease of access for '77.
• Now standard is custom-level interior trim, featuring choice of leather (back cover) or cloth-and-leather seating surfaces (below) on full foam, deep-contoured bucket seats. Select from black, buckskin, red, smoke gray, blue or brown. Or choose from leather-seat interiors in white with buckskin, red, smoke gray or blue accents.

**THE POWER TEAM.**

Get the one that best suits your personal driving preferences.

• The standard 350 4-barrel V8 is a rugged power plant with impressive low-speed torque for around-town cruising, but also very willing and responsive on the highway.
• The standard wide-ratio 4-Speed manual transmission is great for normal driving. The high numerical first-gear ratio gets you off the mark smartly, while wide-range intermediate ratios provide good torque matching at low and intermediate speeds.

| Engines (5.7 Litre) | Trans-missions | Rear Axle Ratios |
|---|---|---|
| Standard 350 4-barrel V8 Net HP*, 180 @ 4000 RPM Torque, 270 ft.-lbs. @ 2400 RPM | 4-Speed, Wide Ratio, Std.†** | 3.36, Std. 3.08, Avail. |
| | Turbo Hydra-matic, Avail. | 3.08, Std. |
| Available† Special 350 4-barrel V8 Net HP*, 210 @ 5200 RPM Torque, 255 ft.-lbs. @ 3600 RPM | 4-Speed, Wide Ratio, Std.†** | 3.70, Std. 3.55, Avail. |
| | 4-Speed, Close Ratio, Avail.†** | 3.70, Std. 3.55, Avail. |
| | Turbo Hydra-matic, Avail. | 3.55, Std. |

SPECIAL NOTE: California Emission Equipment required for registration in California. In other states, High Altitude Emission Equipment may be required in areas 4,000 or more feet above sea level.
*SAE net (as installed) rating.
†Not available in California.
**High Altitude Emission Equipment.

**CORVETTE ENGINES HAVE:**

• High Energy Ignition, a solid-state system that delivers up to 85% hotter spark to the plugs than a conventional system.
• Hydraulic valve lifters.
• Coolant recovery system to help reduce coolant loss.
• Delcotron generator with solid-state regulator.
• Freedom battery needs no refills. Side terminals eliminate corrosion buildup. 3,500-watt rating.
• Long recommended maintenance intervals: Spark plugs, up to 22,500 miles; engine oil, 6 months or 7,500 miles; oil filter, first 7,500 miles, then every 15,000 miles; chassis lube, 6 months or 7,500 miles; automatic transmission fluid, 60,000 miles. Check details in the Owner's Manual.

**FOR SAFETY AND SECURITY.**

*Occupant protection.*

• Two combination seat and inertia reel shoulder belts for driver (with reminder light and buzzer) and right seat passenger.
• Energy-absorbing steering column.
• Safety steering wheel.
• Passenger-guard door locks.
• Safety door latches and hinges
• Folding seat back latches.
• Energy-absorbing padded i
strument panel.
• Thick-laminate windshield
• Safety armrests.

*Accident prevention.*

• Side marker lights an
flectors.
• Parking lamps that illu
with headlamps.
• Four-way hazard warn
flasher.
• Lane-change feature in direction signal control.
• Backup lights.
• Windshield defrosters, washers and dual-speed wipers.
• Wide-view inside day-night mirror (vinyl-edged, shatter-resistant glass and deflecting support).
• Outside rearview mirror.
• Dual master cylinder brake system with warning light.
• Starter safety switch.

*Anti-theft.*

• Alarm set or deactivated by key lock in left front fender.
• Ignition-key reminder buzzer.
• Steering column lock.

• Head room: 36.2".
Shoulder room: 47.9".
Hip room: 48.8".
Leg room: 52.1".
Luggage space: 7.8 cu.ft.

• Power steering.

• Fully independent front and rear suspension.

• Magic-Mirror acrylic exterior finish.

• Fiberglass body construction can't rust or corrode.

• Wide 15"x8" wheels.
Wheelbase: 98.0".
Overall length: 185.2".
Overall width: 69.0".
Loaded height: 48.0".
Tread: front—58.7", rear—59.5".

• Fuel tank rated capacity: 18 gals.

• New stainless steel covers on mufflers—added protection against corrosion.

• Power disc brakes at all four wheels. Fade-resistant and self-adjusting.

• Protective fiberglass inner fenders, front and rear.

• Corvette-specification, GR70-series, steel-belted radial ply tires.

*Unübersehbar der Hinweis, daß die Corvette nach Kalifornien in modifizierter Ausführung geliefert wird. Aber mit Pferdestärken wird nach wie vor nicht gegeizt.*

Chevrolet
CORVETTE
25

*Links: Ende 1977 erschien dieses Faltblatt mit der 25jährigen Geschichte der Corvette. Rechts die Spezifikation des aktuellen Wagens.*

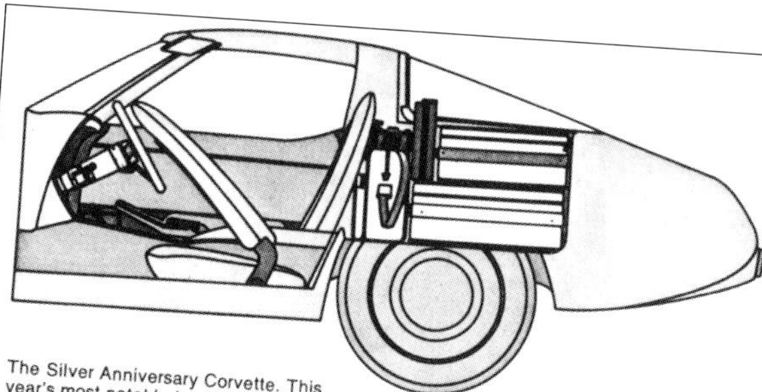

The Silver Anniversary Corvette. This year's most notable improvement is the new fastback roofline and rear window which enhance Corvette's sleek silhouette. Additionally, the new rear window not only allows for a cleaner styling profile, it also improves driver visibility and adds luggage space. There's a roll shade to screen the luggage space. And there's room inside for storage of removable roof panels.

Also new for 1978 is a larger, plastic-lined fuel tank. Capacity has been boosted from 17 to 24 gallons.

The instrument panel has also been restyled and features face-mounted, round instruments which are extremely legible as well as handsome. A new glove box has been added, too.

Corvette performance has been improved with a dual snorkel air cleaner and new low-restriction exhaust system and mufflers on the available L82 350 Cu. In. V8 engine (not available in California). The transverse rear leaf spring in the independent rear suspension has been widened to reduce stress.

Two new transmissions are introduced in 1978. A new Four-Speed manual transmission is standard. A new automatic transmission is available for the L82 350 Cu. In. V8 engine. And in recognition of Corvette's 25th year, Silver Anniversary emblems are mounted on the nose and rear deck.

Corvette continues to offer a long list of standard equipment:

- 4-barrel 350 Cu. In. V8.
- Choice of 4-Speed manual or automatic transmission.
- Independent suspension, front and rear.
- Power steering.
- Reinforced fiberglass body construction that can't rust or corrode.
- Power disc brakes at all four wheels; self-adjusting and fade-resistant.
- Low-profile, steel-belted radial ply tires.
- Wide 15" x 8" wheels.
- Body-colored, energy-absorbing bumpers.
- Flush-mounted, retractable headlights.
- Soft-Ray tinted glass.
- Center console.
- Column-mounted switch that controls turn signals, high and low headlight beams.
- Choice of leather or cloth and leather seat inserts.
- Full foam, deep contoured bucket seats.
- Choice of 10 exterior colors, 7 interior colors.
- Carpeted floor, including tunnel sides and door sills.
- Twin bumper guards, front and rear.
- Removable roof panels.
- Automatic overhead and instrument panel courtesy lights.
- Extensive sound insulation.
- Electric clock.
- Lockable storage compartments.
- Full instrumentation, including speedometer, tachometer, water temperature gage, fuel gage, voltmeter, oil pressure gage, and trip odometer.
- Delco Freedom battery that needs no water.
- High Energy Ignition system.
- Delcotron generator with solid-state voltage regulator.
- Fiberglass inner fenders.

**Long recommended service intervals under normal driving conditions.**

| | |
|---|---|
| Engine oil | 12 months or 7,500 miles |
| Chassis lubrication | 12 months or 7,500 miles |
| Oil filter | First 7,500 miles. Every 15,000 miles thereafter. |
| Spark plugs | Up to 22,500 miles |
| Automatic transmission fluid | 60,000 miles |

**SAFETY FEATURES**

OCCUPANT PROTECTION FEATURES
• Continuous loop seat belt and inertia reel shoulder harness for driver (with reminder light and buzzer) and passenger • Padded sun visors • Energy-absorbing steering column • Passenger guard door locks • Safety door latches and stamped steel hinges • Folding seat-back latches • Energy-absorbing padded instrument panel and front seat-back tops • Thick laminate windshield • Safety armrests • Safety steering wheel.

ACCIDENT PREVENTION FEATURES
• Side marker lights and reflectors • Parking lamps that illuminate with headlamps • Four-way hazard warning flasher • Back-up lights • Lane change feature in direction signal control • Windshield defrosters, washer and dual-speed wipers • Wide-view inside mirror (vinyl-edged, shatter-resistant glass and deflecting support) • Outside rearview mirror • Dual master cylinder brake system with warning light • Starter safety switch.

ANTI-THEFT FEATURES • Anti-theft ignition key reminder buzzer • Anti-theft steering column lock • Anti-theft alarm system with concealed exterior switch, including (for the first time) the roof panels in its circuitry.

**ENGINE AVAILABILITY**

| All states except California | | Transmissions/Rear Axle Ratios | | | |
|---|---|---|---|---|---|
| | | | | Automatic Transmission | |
| Engines | Power Rating† | 4-Speed Manual | 4-Speed Close-Ratio Manual | Below 4,000 Ft. | 4,000 Ft. and Above |
| 350 Cu. In. V8 (Std.) | 185/175▲ | 3.36 | NA | 3.08 | 3.55 |
| 350 Cu. In. V8 (Avail. RPO L82)• | 220 | 3.70/3.36* | 3.70 | 3.55 | NA |
| **California only** | | | | | |
| 350 Cu. In. V8 | 175 | NA | NA | 3.55 | NA |

*†S.A.E. net horsepower as installed. *Available highway ratio.*
*STD.—Standard. NA—Not available. ▲Rating with High Altitude Emission Equipment.*
*•Not available California, Maryland, Florida, Oregon, Washington, also Boston, Chicago, Des Plaines (Ill.), Barrington (Ill.), Grand Rapids (Mi.) and Cook County (Ill.).*
*SPECIAL NOTE: California Emission Equipment required for registration in California. In other states, High Altitude Emission Equipment may be required in areas 4,000 feet or more above sea level.*

GT RADIAL
GOODYEAR
he 25th Anniversary Corvette.
It stands alone today as it has since the summer of 1953, a truly unique
nd finely machined two-seater, America's only true production sports car.
The legend lives on and improves, as legends do, with the passage of time.
The Silver Anniversary Corvette: Twenty-five years in the making, a
we've enjoyed every minute of it.
And now, if you will, a round of applause for the Corvette founding fa
Harley Earl, Zora Arkus-Duntov, Bill Mitchell, Ed Cole…
Available on any Corvette are many Options and Custom Features. Many are illustrated or described in this catalog. Availability often depends on the model and other equipment selected. Your Che

*Der Faltprospekt (Seite 60/61) ließ sich zum Poster ausklappen und enthält auf der Rückseite diese dreifache Dame mit ihrem silberfarbenen Traumwagen Modell 1978. Unten die Titelseite eines 1979er Faltprospektes, der ebenfalls eine Überraschung enthielt – ein gezeichnetes Poster (siehe Seite 64).*

Corvette
The only true American production sport car. An automotive classic that's become a legend among those who know and love fine automobiles. Classic in line, in detail, in capability and in innovative engineering, Corvette has a look, a feel, a way of moving that sets it apart.
Corvette. The pride of Chevrolet. The one and only.
OFFICIAL
Louis Joseph Chevrolet, inventor, mechanic, race car designer, builder and driver, designed and built two successive winners of the Indy 500 in 1920 and 1921.
INDIANAPOLIS SPEEDWAY
A Chevrolet Corvette paced the field for the 1978 Indy 500.
OFFICIAL PACE CAR
1979 Chevrolet Corvette. It's the latest example of the Corvette evolution which began in 1953. Standard equipment includes a 5.7 litre 4 barrel V8, fiberglass body construction reinforced with steel, power disc brakes at all four wheels, independent suspension front and rear and bucket seats for two.
The "Golden Age" of auto racing began in 1911 when Number 32, a Marmon Wasp, won the first Indy 500, the same year Louis Chevrolet designed and built the first Chevrolet car.
32
32
The cockpit. An incredibly designed compartment from which the driver presides over a complete bank of carefully placed instruments. Standard instrumentation includes speedometer with trip odometer, tachometer, water temperature gage, oil pressure gage, voltmeter, fuel gage and clock. Space is also provided for available CB radio, stereo system, air conditioning and automatic speed control.
'79 corvette
Retractable dual headlights wink out of sight by day.
Passenger seat folds flat on cushion for easy access to rear storage.
Tilt-Telescop
is an often-
option
Concealed

*Die Zusammenstellung diverser Extras zu einem ,,persönlichen'' Wagen ließ die GM-Texter zur Aussage verleiten: Bauen Sie sich Ihre eigene Corvette!*

# Build your own Corvette.

(See your dealer for current ordering information and specific model/option availability.)

**COLORS**

Exterior color ________

Interior color† ________

Upholstery ________

**ENGINES**

(See Power Teams chart for details on engine and transmission availability.)

- ☐ 5.7 Litre 4-Bbl. V8 (350 Cu. In.) (Std.)
- ☐ 5.7 Litre 4-Bbl. V8 (350 Cu. In.) L82 (Avail.)*

**TRANSMISSIONS**

- ☐ Four-Speed manual*
- ☐ Four-Speed manual close-ratio*
- ☐ Automatic

**AVAILABLE OPTIONS**

- ☐ Air conditioning.
- ☐ Removable tinted glass roof panels.
- ☐ Tilt-Telescopic steering wheel.
- ☐ Automatic speed control.
- ☐ Electric rear window defogger.
- ☐ Power windows.
- ☐ Power windows and door locks.
- ☐ Sport mirrors, LH remote and RH manual.
- ☐ Convenience Group (includes time-delay dome and courtesy lights, headlight warning buzzer, underhood light, low fuel warning light, color-keyed floor mats, intermittent windshield wipers and RH visor vanity mirror).

**SOUND EQUIPMENT**

- ☐ AM/FM stereo/Citizens Band radio with tri-band power antenna.
- ☐ AM/FM stereo radio.
- ☐ AM/FM stereo radio with 8-track stereo tape.
- ☐ AM/FM stereo radio with stereo cassette tape.
- ☐ Power antenna.
- ☐ Dual rear speakers.

**TIRES**

- ☐ P225/70R-15 steel-belted radial ply white-lettered tires.
- ☐ P255/60R-15 aramid-belted radial ply white-lettered tires.

**MISCELLANEOUS**

- ☐ Aluminum wheels.
- ☐ Gymkhana suspension.
- ☐ Chassis equipment — trailering.
- ☐ Heavy-duty battery.
- ☐ Highway axle ratio.
- ☐ Heavy-duty shock absorbers.

**Not available in California or with High Altitude Emission Equipment.*

† Leather or cloth and leather interior colors available: light beige, dark blue, dark green, oyster. Leather only: black, red.

## A word about assembly, components and optional equipment in these Chevrolets.

The Chevrolets described in this brochure are assembled at facilities of General Motors Corporation operated by the GM Assembly Division. These vehicles incorporate thousands of different components produced by various divisions of General Motors and by various suppliers to General Motors. From time to time during the manufacturing process, it may be necessary, in order to meet public demand for particular vehicles or equipment, or to meet federally mandated emissions, safety and fuel economy requirements, or for other reasons, to produce Chevrolet products with different components or differently sourced components than initially scheduled. All such components have been approved for use in Chevrolet products and will provide the quality performance associated with the Chevrolet name.

With respect to extra cost optional equipment, make certain you specify the type of equipment you desire on your vehicle when ordering it from your dealer. Some options may be unavailable when your car is built. Your dealer receives advice regarding current availability of options. You may ask the dealer for this information. GM also requests the dealer to advise you if an option you ordered is unavailable. We suggest you verify that your car includes the optional equipment you ordered or, if there are changes, that they are acceptable to you.

*Seiten aus einem 1981er Corvette-Prospekt.*

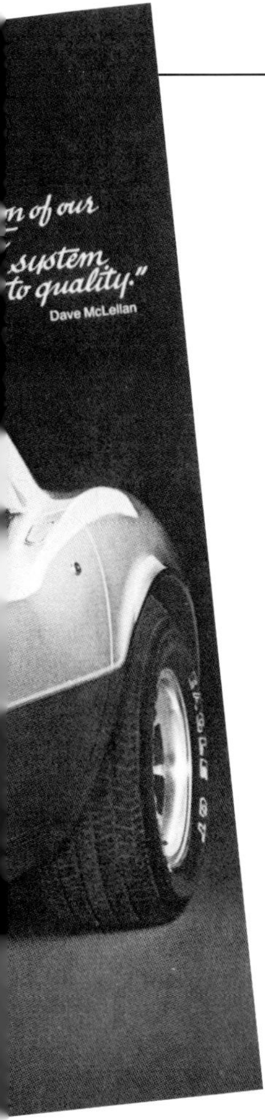


Corvette is a rolling showcase of new technology, new materials and new thinking—all part of an evolutionary process that continues year after year. Adherence to the principle of engineering excellence has been applied to all areas, from the front bumpers to the aircraft-style cockpit to the rear suspension system. It is these kinds of developments—detailed here—that make Corvette what it is today: one of the most renowned two-seat sports cars in the world.

For 1981, Corvette engineers trod where none have gone before by developing the world's first and only fiberglass-reinforced composite automotive spring (used with automatic transmission only). It even won the Society of Plastic Engineers Grand Award. This monoleaf rear spring—at 8 lbs.—replaces a 41-lb. steel multileaf spring. And it's not just lighter, it's more efficient. It can store six times the strain-energy per unit weight than a comparable weight of spring steel. This is the lightest possible leaf spring application to be found in any automotive suspension. And steel interleaf friction is eliminated. This fiberglass spring supports the weight of the car that rests on the rear wheels and provides suspension compliance. Wheel position is controlled by a system of links and pivots.

Attention to detail for 1981 includes other engineering developments such as magnesium valve rocker covers and stainless steel exhaust manifolds. And an

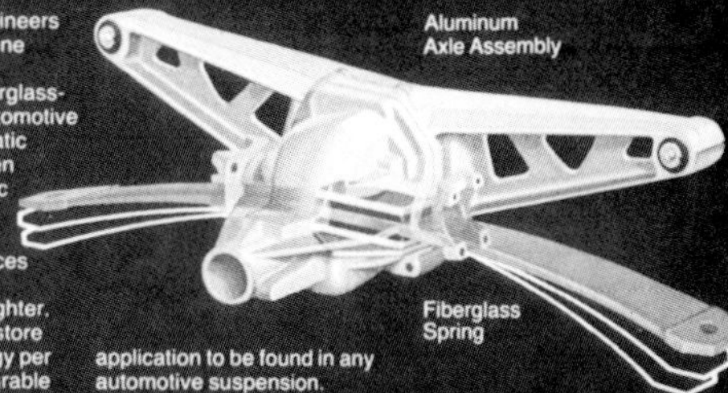


*"We critique Corvette with the same engineering objectivity we'd use to evaluate a military aircraft: What is Corvette's mission? How well does it carry out that mission?"*

Dave McLellan,
Chief Engineer, Corvette

1982 CORVETTE
Chevy makes good things happen.
corvette
Chevrolet
te Coupe.

# 1982 Corvette

## This is a whole new kind of Corvette.

It is still the road-rapacious sports car that's always been capable of inhaling great chunks of tormented 2-lane tarmac and leaving you exhilarated even after long, 1000-mile days in the saddle. It's still the time and space machine it's always been.

But there's a new civility now, predicated in large part on two engineering advances for 1982. Cross-fire fuel injection and a 4-speed automatic transmission with overdrive fourth gear so good in every aspect of driving usage that the manual transmission has been deleted.

There's a new quality about Corvette too. A space age standard of fit and finish quality that is rigidly imposed and then double-checked by workers and inspectors at Corvette's new 1-million-square-foot plant in Bowling Green, Kentucky.

And for this year, a unique and distinguished Corvette . . . the special "Collector Edition".

This is the Corvette to covet. From its raised white-lettered Eagle GT tires to its frameless hatchback, the "Collector Edition" is, we believe, the finest all-around 2-seater in the world.

*Corvette Coupe.*

*Frameless glass-hatch design with remote release is standard on the "Collector Edition" Hatchback Coupe.*

*Aircraft-style centre console and full gauge instrumentation are standard.*
*NOTE: Speedometers in Canada have major scale in km/h, secondary in mph and odometer in kilometres.*

*Corvette standard interior.*

*Einfach in Druck und Gestaltung ist dieser Prospekt. Aber er zeigt alle Einzelheiten.*

## Corvette Major Standard Equipment

| EXTERIOR | Corvette Sport Coupe | Corvette Collector Edition |
|---|---|---|
| Anti-theft alarm system with starter-interrupt feature | S | S |
| Front fender louvres | S | S |
| Front cornering lamps | S | S |
| Specific "Collector Edition" emblems – front, sides, rear | NA | S |
| Power-operated retractable headlamps | S | S |
| Quad headlamps with halogen hi-beam inner units | S | S |
| Dual remote-controlled outside sport mirrors | S | S |
| Tinted glass | S | S |
| Special silver-beige metallic finish | NA | S |
| Removable roof panels with solar screening | EC | S |
| Steel wheels with centre hub and trim rings | S | NA |
| Finned aluminum wheels | NA | S |
| Frameless rear hatch glass with remote release | NA | S |
| Body-colour front bumper with integral air dam | S | S |
| Energy-absorbing bumper systems | S | S |
| Corrosion-resistant steel-reinforced fiberglass body | S | S |
| Concealed dual-speed wipers with integral washers and wiper arms | S | S |
| **INTERIOR** | | |
| Leather-trimmed vinyl or all-cloth bucket seats | S | NA |
| Full-leather bucket seats | NA | S |
| Soft-padded and carpeted door panels | S | S |
| Moulded shell seats with high-pivot folding backs | S | S |
| Power windows | S | S |
| Electric rear window defroster | S | S |
| AM/FM push-button radio with dual front speakers and mast antenna | S | S |
| Air conditioning | S | S |
| Time-delay dome and courtesy lights | S | S |
| Headlamp-on reminder | S | S |
| Low-fuel warning lamp | S | S |
| Illuminated RH visor vanity mirror | S | S |
| Leather-wrapped steering wheel rim | S | S |
| Tilt-Telescopic steering wheel and column | S | S |
| Glove compartment lock and lamp | S | S |
| Intermittent windshield wipers | S | S |
| 7,000-RPM electronic tachometer | S | S |
| Voltmeter, temperature, fuel and oil pressure gauges | S | S |
| Quartz analog electric clock with sweep-second hand | S | S |

| | Corvette Sport Coupe | Corvette Collector Edition |
|---|---|---|
| Cigarette lighter and ashtray | S | S |
| Resettable trip odometer | S | S |
| Centre console with shifter, coin tray, window, mirror and air conditioning controls | S | S |
| Day/night rearview mirror | S | S |
| Deep-twist floor and stowage area carpet | S | S |
| Acoustical insulation package | S | S |
| Luggage compartment concealment shade | S | S |
| **POWER TEAMS/CHASSIS/MECHANICAL** | | |
| Computer Command Control | S | S |
| 5.7 Litre CFI V8 engine (Cross-Fire Injection) | S | S |
| Automatic transmission with overdrive fourth gear | S | S |
| Aluminum intake manifold with tuned runners | S | S |
| Stainless steel exhaust manifolds and free-flow mufflers | S | S |
| Hydraulic valve lifters and exhaust valve rotators | S | S |
| Magnesium valve rocker covers | S | S |
| Black wrinkle-finish air cleaner | S | S |
| Hood outside-air duct | S | S |
| Electric in-tank fuel pump | S | S |
| Low-drag engine primary cooling fan | S | S |
| Auxiliary electric engine cooling fan | S | S |
| High Energy Ignition system | S | S |
| Freedom II battery with sealed side terminals | S | S |
| Power steering with ball-race gear | S | S |
| Power disc brakes at all four wheels | S | S |
| Exclusive monoleaf fiberglass rear spring design | S | S |
| New driveshaft seals and splash shields | S | S |
| New, larger drive shaft universal joints | S | S |
| Sturdy frame structure with corrosion-resistant coating | S | S |
| Fully independent four-wheel suspension | S | S |
| Delcotron generator with built-in solid-state regulator | S | S |
| P225/70R-15B steel-belted radial ply tires | S | NA |
| P225/60R-15B steel-belted radial ply white-lettered tires | EC | S |
| Compact spare tire | S | S |
| Side-lift jack | S | S |
| Underhood lamp | S | S |

S–Standard EC–Optional at Extra Cost NA–Not Available

### Options

Listed below are some of the available equipment choices to help you customize your 1982 Corvette. For other optional equipment and their availability, see your Chevrolet dealer.

• Electric twin remote control sport mirrors • Roof panel carrier • Power door lock system • Delco radios: AM/FM stereo radio • Electronically tuned stereo radio with 8-track tape player • Electronically tuned stereo radio with cassette tape player • Electronically tuned stereo radio with Citizens Band and cassette tape • Power antenna (included with Citizens Band radio) • Removable glass roof panels • Automatic speed control with 'resume speed' feature • Gymkhana suspension includes rear stabilizer and bushings, higher-rate springs and special shock absorbers • Tires: P225/70R-15 or P225/60R-15 steel-belted radial ply white-lettered • Six-way power driver's seat • Aluminum wheels • Custom two-tone exterior paint.

### Exterior Colours

Choose from 11 high-lustre acrylic finishes or several available Two-Tone combinations. Colours: White, Silver Metallic, Black, Silver Blue Metallic, Dark Blue Metallic, Silver Green Metallic, Bright Blue Metallic, Gold Metallic, Charcoal Metallic, Red, Dark Claret Metallic.

Corvette "Collector Edition" features an exclusive Silver Beige Metallic exterior finish.

### Interiors

Corvette is available in a choice of four cloth interiors: Silver Gray, Dark Blue, Camel and Dark Red; or select from six available leather trims: Silver Gray, Charcoal, Dark Blue, Green, Camel or Dark Red.

Corvette "Collector Edition": features full leather seats in an exclusive Silver Beige multi-tone trim.

### Safety Features

OCCUPANT PROTECTION
Manual lap/shoulder belts with push-button buckles for driver and right front passenger (driver's side includes visual and audible warning system) • Energy-absorbing steering column • Passenger guard door locks • Safety door latches and stamped steel hinges • Folding seat back latches • Energy-absorbing padded instrument panel with anti-reflective upper surface • Laminated windshield/tempered side and rear glass • Safety armrests • Identification symbols for controls and displays.

ACCIDENT AVOIDANCE
Side marker lights and reflectors • Parking lamps that illuminate with headlamps • Four-way hazard warning flasher • Backup lights • Lane change feature in direction signal control • Windshield defrosters, washer and dual-speed wipers with time-delay feature • Vinyl-edged inside mirror • Dual outside rearview mirrors • Dual master cylinder brake system with warning light • Starter safety switch.

ANTI-THEFT
Audible anti-theft ignition key reminder • Anti-theft steering column lock • Anti-theft audio alarm system with starter-interrupt feature • Inside hood release.

### A word about assembly, components and optional equipment in these vehicles.

The vehicles described in this brochure are assembled at various Divisions of General Motors Corporation. These vehicles incorporate thousands of different components produced by General Motors of Canada Limited and related companies and their suppliers. From time to time during the manufacturing process, it may be necessary, in order to meet public demand for particular vehicles or equipment, or to meet federally mandated emission and safety requirements or fuel economy guidelines, or for other reasons, to produce vehicles with different components or differently sourced components than initially scheduled. All such components have been approved for use in our products by General Motors of Canada Limited or related companies and will provide the quality performance associated with our name.

Please note that certain optional equipment can be ordered only in conjunction with other optional equipment. Also, upgrading from specified standard equipment may be necessary when certain optional equipment is ordered. Ask your sales representative for details.

With respect to extra cost optional equipment, make certain you specify the type of equipment you desire on your vehicle when ordering it from your dealer. Some options may be unavailable when your car is built. Your dealer receives advice regarding current availability of options. You may ask the dealer for this information. GM also requests the dealer to advise you if an option you ordered is unavailable. We suggest you verify that your car includes the optional equipment you ordered or, if there are changes, that they are acceptable to you.

All illustrations and specifications contained in this catalogue are based on the latest product information available at time of publication approval. The right is reserved to make changes at any time, without notice, in prices, colours, materials, equipment, specifications and models, and to discontinue models. Some of the equipment shown or described throughout this catalogue is available at extra cost. Check with your Chevrolet dealer for complete information.

General Motors of Canada Limited
Oshawa, Ontario

Printed in Canada November 1981

*Technische Einzelheiten der Corvette Jahrgang 1982. Auf die „Collector Edition" wird ausführlich eingegangen.*

*Multi-tone full leather seats highlight the "Collector Edition" special interior.*

*Corvette Collector Edition Hatchback Coupe.*

**Exterior Dimensions: Millimetres (Inches)**

| | | |
|---|---|---|
| Wheelbase | 2489 | (98.0) |
| Length (overall) | 4707 | (185.3) |
| Width (overall) | 1753 | (69.0) |
| Height (loaded) | 1219 | (48.0) |
| Tread – front | 1491 | (58.7) |
| Tread – rear | 1511 | (59.5) |
| Fuel tank capacity: Litres (gals) | 90.8 | (20) |

**Interior Dimensions: Millimetres (Inches)**

| | | |
|---|---|---|
| Head room – front | 919 | (36.2) |
| Leg room – front | 1069 | (42.1) |
| Hip room – front | 1267 | (49.9) |
| Shoulder room – front | 1206 | (47.5) |
| Usable trunk capacity: Litres (cu. ft.) | 234 | (8.4) |

**Corvette Power Teams**

| Engine | Ordering Code | Displacement (cu. in.) | Engine Availability | Transmission Availability |
|---|---|---|---|---|
| 5.7 Litre Cross-fire Injection V8(A) | L83 | 350 | Std. | Std. 4-speed automatic with overdrive |

Std. = Standard (A) = Produced by GM – Chevrolet Motor Division, U.S.A.
Computer Command Control System (NB1) standard.
Caution: The use of leaded fuel will cause the Computer Command Control System to malfunction.

**See your Chevrolet salesperson for Transport Canada approved fuel consumption estimates for all Corvette models.**

**A WORD ABOUT ENGINES**
Some Chevrolets are equipped with engines produced by other GM division, subsidiaries, or affiliated companies worldwide. See your dealer for details.

*Exportorientiert: Angaben in metrischen Maßen! Das Motorenangebot hat sich drastisch verringert . . .*

COMMITMENT

The fact that the first chapter in this Corvette brochure concerns the final chapter in the Corvette building process is in itself significant. We are committed to excellence in the production of the product, in order to achieve excellence in the product.

A good example of the totality of the Corvette commitment is the scene pictured here – "The Morning Audit." Each morning, salaried and hourly employes m[…] examine cars in detail a[…] search for ways to assu[…] enhance the quality of C[…]

It is typical of the atten[…] detail in evidence throu[…] the new Corvette plant […]

The Corvette plant, Bowling Green, Kentucky

10

Eine neue Generation: Corvette-Katalog für den Modelljahrgang 1983.

# Alles neu – macht GM 1983

In gänzlich neuer Aufmachung zeigte sich die Corvette im Modell-Jahrgang 1983. Geblieben war lediglich das V8-Injection-Aggregat mit 350 SAE-PS. Von seinem ursprünglichen „Biß" hatte diese Corvette viel verloren, doch inzwischen war eine neue Automobil-Generation herangewachsen, die ganz andere Maßstäbe anlegte . . .

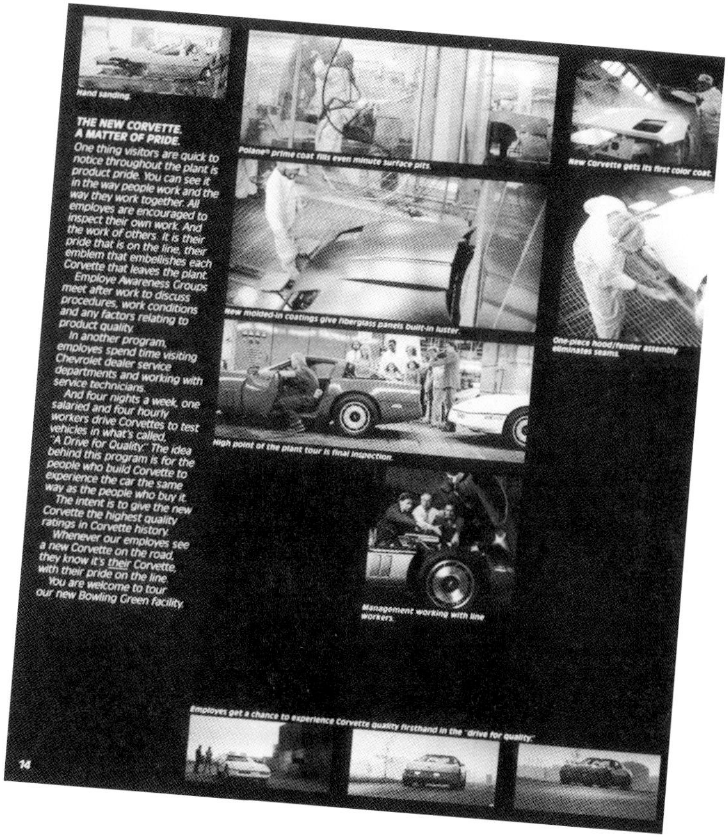

Hand sanding.

**THE NEW CORVETTE. A MATTER OF PRIDE.**

*One thing visitors are quick to notice throughout the plant is product pride. You can see it in the way people work and the way they work together. All employes are encouraged to inspect their own work. And the work of others. It is their pride that is on the line, their emblem that embellishes each Corvette that leaves the plant.*

*Employe Awareness Groups meet after work to discuss procedures, work conditions and any factors relating to product quality.*

*In another program, employes spend time visiting Chevrolet dealer service departments and working with service technicians.*

*And four nights a week, one salaried and four hourly workers drive Corvettes to test vehicles in what's called, "A Drive for Quality." The idea behind this program is for the people who build Corvette to experience the car the same way as the people who buy it.*

*The intent is to give the new Corvette the highest quality ratings in Corvette history.*

*Whenever our employes see a new Corvette on the road, they know it's their Corvette, with their pride on the line.*

*You are welcome to tour our new Bowling Green facility.*

Polane® prime coat fills even minute surface pits.

New Corvette gets its first color coat.

New molded-in coatings give fiberglass panels built-in luster.

One-piece hood/fender assembly eliminates seams.

High point of the plant tour is final inspection.

Management working with line workers.

Employes get a chance to experience Corvette quality firsthand in the "drive for quality."

14

You'd expect the new Corvette to brake handily and it does. With a Girlock four-wheel disc brake system, the new Corvette can decelerate from its top speed as high as 1.0g. Girlock is internationally respected for high performance and racing brake expertise. The low-drag, aluminum caliper design for the new Corvette effects a considerable weight savings over previous Corvette four-wheel disc systems, which helps to reduce unsprung mass. The resulting vehicle dynamics are impressive.

With the standard suspension and the optional Goodyear Eagle VR tires, lateral acceleration is rated at 0.9g on our skidpad, in the hands of a professional driver.

1.0g is equal to the pull of gravity. Cornering or braking at 1.0g means we are encountering a horizontal component (or force) equal in pull to the gravity vector (or the full weight of the car). With Corvette's 0.9g reading, the lateral force is equal to 90% of the weight of the car.

There is an optional suspension, the Z51 Performance Handling Package. It generates 0.95g under lateral acceleration on GM's skidpad. Readings in this range are remarkable, yet Corvette is still a comf
able car to drive.

The Z51 incorporates 1
9½" rear wheels, larger
stabilizer bars, 25mm fro
23mm rear. Different bus
are used, allowing low defle
while cornering. There are

36

Man gibt sich wieder aggressiv – und sei es durch ungewöhnliche Fotowinkel – aber auch ausführlich im Detail.

of the proven and efficient racing engineering practices. The use of forged aluminum suspension components reduced unsprung mass by nearly 30%, which contributes to wheel control, ride and road holding. The payoff is that the new Corvette has demonstrated remarkable directional stability and transient response in surface testing on the GM proving grounds.

The new Corvette. A true world-class sports car.

ences, too, in the stiffness fiberglass monoleaf s at both front and rear the valving of the shocks. 3.0:1 ratio, power-assisted nd-pinion steering of 51 package makes for response to driver control

suspension of the new tte is a totally integrated ge which reflects many

37

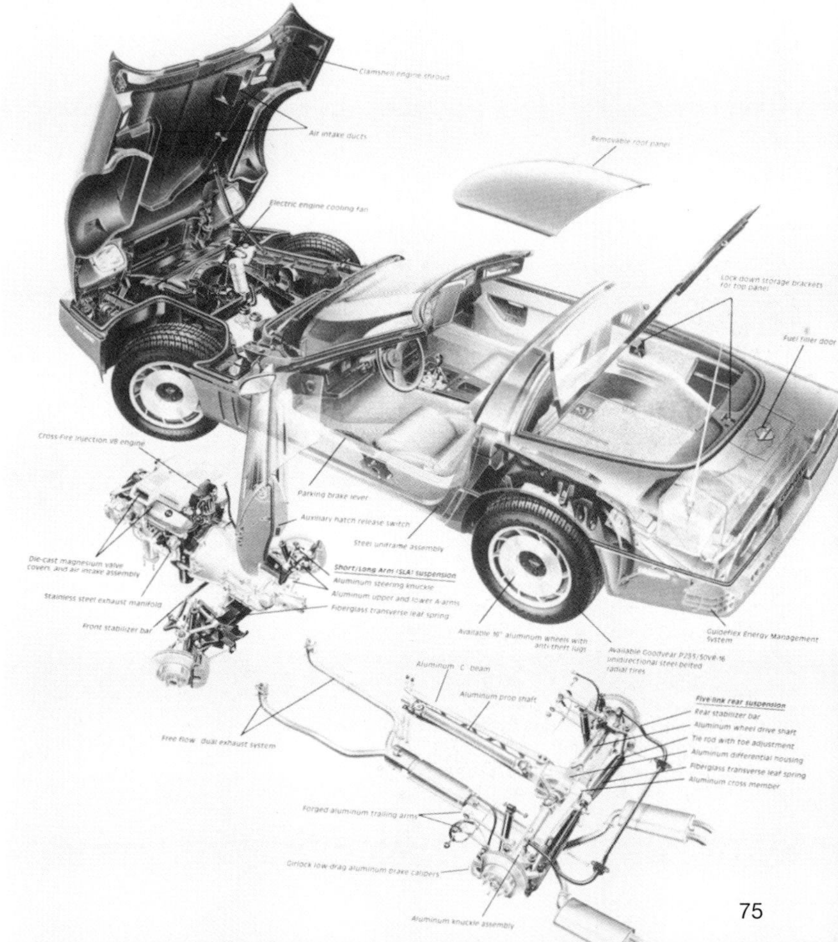


*Inzwischen wurde der Targa-Bügel zum Styling-Element.*

*Katalog des Modells 1986.*

CARSON WELLS
HOTEL

*Der schönste Corvette-Katalog, den es je gab – mit 44 Seiten Inhalt und eingelegten Datenblättern. Auch die Fotos sind von ausgezeichneter Qualität.*

NEEDLES, SOUTH DAKOTA. NAMED FOR THE SHARP PEAKS THAT JUT UP FROM THE GRASSES OF THE GREAT PLAINS. HOME OF PRAIRIE DOGS AND BISON. HOST, BRIEFLY, TO CORVETTE.

AMERICA THE VAST, WHERE ROADS THAT STRETCH STRAIGHT TO THE HORIZON ARE AN IDEAL AND INVITING PROVING GROUND FOR THE POWER THAT PULSES THROUGH CORVETTE.

You'd think the sky is the sky everywhere. But people who've grown up under unbroken western vistas feel hemmed in when they travel eastward. And those who like their sky in bits and pieces gaze 360°, shiver a little and say, "There's too much sky," when they go out West.

In the Corvette Convertible you do more than travel under the sky. You become part of it. With nothing between you and the changing face of America... the pungence of freshly turned farmland... the wind blowing from Idaho to Indiana... the Rockies thrust up along the spine of the continent ...deserts spread over endless arid miles.

Before long the car assumes the aura of its occupant. Temperature and airflow set to particular preferences. Favorite traveling music modulated precisely. A sweater tossed aside.

When you climb aboard after a brief rest stop, you simply assume everything will be as you left it. After all, you and the Corvette form a private world.

Private but never anonymous. Not in this car. From urban streetside to rural truck stop to western national park, the roadster draws a crowd. Murmurs of approval follow. Children dash excitedly from nose to spoiler as young couples grip hands and promise each other, "Someday." Everybody recognizes the Convertible. Most want to touch it as if a piece of The Dream might rub off. And as the machine speeds into the distance, admiration for its beauty mixes with the wish to be going anywhere in the envious glances of those left behind.

Maybe that's because America and cars

ESCORT
VALVOLINE
JEN-SEN
BFGoodrich
duraflame
Hella

*GOODYEAR VR50 TIRES*
The standard tire and wheel combination features P255/50VR 16 Goodyear unidirectional steel-belted radial tires mounted on 16" aluminum alloy wheels with functional turbine-blade design and anti-theft nuts.

*ROOF PANEL STORAGE (COUPE)*
The removable one-piece roof panel combines the security of a closed car with the exhilarating effects of an open roadster. There is no T-bar. The roof is removed using a ratchet wrench designed for this application. The panel may then be stored within the car in a lockdown position.

*HALOGEN FOG LAMPS*
The integral halogen fog lamps make driving under adverse conditions easier. Operation is independent of the headlamps.

*CLOTH BUCKET SEATS*
The high-back cloth bucket seats are contoured to provide the feeling of individualized fit and comfort. Manual back angle adjustment is offered. Other interior features include a leather-wrapped steering wheel, dual rear lockable storage compartments,* lighted visor vanity mirror.

*TILT-TELESCOPIC STEERING WHEEL*
Corvette adjusts to your driving style with a steering wheel that tilts and telescopes.

*4-SPEED AUTOMATIC TRANSMISSION*
Four-speed automatic overdrive transmission with lockup torque converter contributes to impressive performance plus low-RPM highway cruising.

*4-SPEED MANUAL TRANSMISSION*
If you prefer, choose the smooth-shifting manual 4-speed, with an electronic automatic overdrive feature in 2nd, 3rd and 4th gear, which is available at no extra cost. Gear ratios are chosen to deliver exceptional performance.

*CORVETTE NEWS*
A three-year complimentary subscription goes to every Corvette buyer. This publication keeps owners up to date on their favorite sports car.

*Coupe only.



CORVETTE'S AIR CONDITIONING ALLOWS YOU TO MAINTAIN A PERFECT DRIVING ENVIRONMENT.

RETRACTABLE HEADLAMPS FLIP FORWARD 162.5 DEGREES FROM BENEATH THE HOOD.

VATS EMPLOYS A DECODER AND IGNITION KEY WITH A PELLET OF SPECIFIED RESISTANCE.

GOODYEAR EAGLE GT P255/50VR 16 UNIDIRECTIONAL RADIAL TIRES.

THE REMOVABLE ONE-PIECE ROOF LIFT-OFF PANEL STORES IN A LOCKDOWN POSITION IN THE REAR COMPARTMENT, OPENING CORVETTE UP TO SUN AND STARS.

*Motorsport: gewiß kein vorder gründiges Corvette-Thema – aber sehr fotogen . . .*

*Das Herz der '86er Corvette – der 5,74-Liter-V8-Motor mit gemäßigten 208 PS. Kraftstoff-Einspritzung ist mittlerweile zur Selbstverständlichkeit geworden.*

# CORVETTE, A GENERATION:
## 25 YEARS OF MEN, MACHINES, AND MEMORIES.

1953 – 1955
Januar 1953: Harley Earl präsentiert die Corvette auf der New Yorker Motorama-Show. Im Mai 1953 ist der Wagen reif für die Serienfertigung, vier Wochen später erscheint das erste Exemplar einer 300-Stück-Serie. Im September 1954 ist auch ein V8-Motor erhältlich. Der Roadster erhält im Herbst 1955 Kurbelscheiben und ein Hardtop.

1956 – 1957
Im September 1956 wird die Corvette mit Benzineinspritzung versehen. Für das Modelljahr 1958 bekommt der Wagen Doppelscheinwerfer und mehr Chrom.

## Corvette – Auto einer Generation

1958 – 1960
GM-Designer Bill Mitchell arbeitet an einem Modell mit neuer Karosserie; es wird später als Sting Ray berühmt. Im Sommer 1960 wird die 10 000. Corvette des laufenden Modelljahrgangs hergestellt. Mit neuem Heck präsentiert man den 1961er Wagen.

1961 – 1962
Im Februar 1961 gibt der Sting Ray als Show Car in Chicago sein Debüt. Im Dezember 1962 testet man Scheibenbremsen und stärkere V8-Motoren.

1963 – 1967
Ab Ende 1963 weist das Heckfenster des Coupés keinen Steg mehr auf. Ein Jahr später bekommt die Corvette rungsum Scheibenbremsen. 6,5- und 7-Liter-Motoren werden angeboten. Das Modell 1966 wird stilistisch überarbeitet; die Motoren haben keine Einspritzung mehr. Im April 1967 kommt eine Hochleistungs-Version (L88) heraus mit 435 PS (und ohne Heizungsanlage).

1968 – 1977
Stark überarbeitetes Karosserie- und Interieur-Design, ähnlich dem Show Car „Mako Shark II". Im November 1969 verläßt die 250 000. Corvette das Werk in St. Louis. Im Februar 1970 stellt GM einen neuen 7,5-Liter-Motor – bis 460 PS stark – vor. Karosserieretuschen im Herbst 1973. Große Feierlichkeiten bei General Motors: Im März 1977 wird die 500 000. Corvette gebaut.

(aus einem Prospekt im Frühjahr 1978)

CORVETTE

# Bücher für den Corvette-Liebhaber

**Chevrolet – A History from 1911** – von Beverly Rae Kimes und Robert Ackerson. Die Corvette findet in diesem Geschichtsbuch ihren gebührenden Platz; vor allem sind die Farbfotos exzellent. Das Buch umfaßt insgesamt 200 Seiten mit 160 meist farbigen Fotos.

**Corvette, an American Legend** – vom Verlag Automobile Quarterly. Eines der ausführlichsten und bestgemachten Bücher zum Thema Corvette. Insbesondere werden Wagen der Baujahre 1953 bis 1967 eines bestimmten Wettbewerbs (Bloomington) behandelt, die absolut originalgetreu sind. 184 Seiten, 200 sehr gute Bilder in Farbe. Englischer Text.

**Corvette** – von B. Coleman. Eine Zusammenfassung der Corvette-Historie auf 100 Seiten (1953 – 1983) mit 100 Fotos, die meisten farbig. Ein guter Überblick (engl. Text) zur Schnell-Information.

**Classic Corvettes in a Nutshell** – von John F. Ball. Kaufberatung vom Fachmann, voller nützlicher Tips (engl. Sprache). 102 Seiten, 22 Abbildungen.

**Corvette 1953 – 1967, A Source Book** – von Arthur A.C. Steffen. Auf 144 Seiten Faksimile-Drucke von alten Werbemitteln und Inseraten, auch Händler-Informationen von General Motors. Eine interessante Quelle authentischer Unterlagen „aus der Zeit" mit über 300 Abbildungen.

**Corvette – the Legend lives on.** Die Fortsetzung des Bloomington-Themas für Wagen der Baujahre 1968 bis 1982, von Automobile Quarterly (siehe oben). 200 Seiten, 250 Farbfotos.

**Early Chevrolet Corvettes** – von C. Falconer. Ein Band der Auto-History-Serie (engl. Sprache). Alle Sechs- und Achtzylinder 1953 – 1967 werden hier beschrieben. 136 Seiten, 60 Schwarzweiß-Abbildungen, 8 Farbseiten.

**Corvette Performance** – alles über die Entwicklung dieses Wagens als Sport- und Rennfahrzeug (engl. Text). Renn- und Testberichte, Clubinformationen. 458 Abbildungen auf 180 Seiten.

**Corvette, Amerikas Sportwagen** – von Jay Koblenz. Ein in deutscher Sprache im Serag-Verlag erschienenes Buch von 256 Seiten Umfang der gesamten Geschichte des amerikanischen Zweisitzers. 250 Schwarzweiß- und 145 Farbaufnahmen. Ein lesenswertes und gut aufgemachtes Buch für den Enthusiasten.

**The Complete Book of the Corvette** – von Richard M. Langworth. Der prominente amerikanische Automobilhistoriker schrieb diesen hervorragend aufgemachten Band, der alle Modelle bis einschließlich 1987 vorstellt. 320 Seiten, über 400 vorwiegend farbige Abbildungen.

**Corvette 1953 – 1980 Identification and Number Codes.** Alle Chassisnummern und Motordaten, Getriebe-Spezifikationen und so fort für die Baujahre von 1953 bis 1980, auch Lackierungen (ab 1963) und Extras. Für den Besitzer einer Corvette – der es ganz genau wissen will – ein wertvolles Nachschlagewerk.

**Corvette Restoration Handbook.** Eine Restaurierungsanleitung für den Amateur, sehr gut abgefaßt, mit mehr als 300 Abbildungen (engl. Text) auf 248 Seiten. Alle Modelle bis 1967 finden Berücksichtigung.

**Corvette Sting Ray** – von Robert Ackerson. Ein Super-Profile-Band mit ausführlicher Typengeschichte, Restaurierungstips und Literatur-Verzeichnis. 56 Seiten, 20 farbige Abbildungen.

**Corvette, Sports Car for America** – von M.B. Antonick. Eine gut aufgemachte Rückblende auf die Entwicklungsgeschichte der Corvette von 1953 bis 1955, mit zeitgenössischen Werbemitteln und anderen Abbildungen. 240 Seiten, 200 Schwarzweiß- und 55 Farbfotos. Englischer Text. Der Autor hat auch ein weiteres Buch über die Corvette verfaßt, das nur spezielle Wettbewerbswagen behandelt.

**The Complete Corvette Restoration & Technical Guide** – von N. Adams. Ein teures, aber unentbehrliches Werk für den Restaurator. Mit 1400 Abbildungen auf 424 Seiten. Gründlicher und informativer geht's nicht! Englischer Text.

**The Road Corvette** – von R. Miller und G. Embree. Über 100 Fotos aller Modelle und Derivate von 1953 bis 1973 enthält dieses Buch, das zu den Klassikern der Corvette-Literatur gehört. Es hat 320 Seiten und enthält unter anderem sämtliche technischen Daten im Detail.

**Corvette Body Repair Guide** – von Robert J. Schiro. Für die Reparatur der Kunststoff-Karosserie ein wertvoller Helfer. Alle Vorgänge werden in Bildern gezeigt (250 Abbildungen, auch in Farbe). Das Buch ist spiralgebunden und umfaßt 200 Seiten.

(Es gibt noch etliche weitere Publikationen, sämtlich aus amerikanischen Verlagen, sowie Buyers Guides und Straßentest-Reprints. Die Automobilbuch-Versandhäuser und -Fachgeschäfte bieten eine umfassende Auswahl an einschlägiger Corvette-Literatur.)

*Shown on Cover: Corvette Sting Ray Sport Coupe in Tuxedo Black*

*Corvette Sting Ray Convertible in Riverside Red*

# NEW GRACE AND ELEGANCE SILHOUETTE AMERICA'S SPORTS CAR

Corvette steps out smartly with an exhilarating new look for '63. A freshly elegant look that promises to lift the spirits of any buff who takes the wheel. It's the Corvette Sting Ray! Two sparkling new models, both pointing the way to a classic new concept in American sports car design.

Catch an eyeful of the Corvette newcomer, the exciting Sport Coupe. Graceful body panels and compound-curved side windows emphasize the miles-ahead aerodynamic design. Power-operated, retractable headlamps rotate out of sight to blend into the unobstructed hood line. Doors extend upward into the roof, adding armchair ease and comfort to sliding in and out. This is the Corvette Sting Ray Sport Coupe. Clean, taut, fresh in every detail.

You'll find the Convertible is jauntier than ever in '63. Uncluttered lines accented front and rear by trim wrap-around bumpers. The subtle contour of the smoothly molded rear deck. A gleaming aluminum grille, retractable head-lamps and a sweeping hood-length windsplit. Top up or down, graceful streamlining is the goal. And the Corvette Sting Ray Convertible achieves it.

# Die Corvette in der Presse

Im August 1953 erschienen in der amerikanischen Motorpresse die ersten Berichte über die neue Corvette. Sie gingen über eine Modellbeschreibung kaum hinaus – Straßentests wurden nicht vor dem Frühjahr 1954 veröffentlicht. In **Road & Track** las sich ein erster Eindruck so:

,,Schon seit 1948 erhoben wir immer wieder die Frage nach einem amerikanischen Sportwagen. Jetzt haben wir ihn in Gestalt der Chevrolet Corvette. Allerdings stellten wir anläßlich einer Testfahrt nach Chicago und New York fest, daß dieses Auto bei Sportwagen-Enthusiasten noch kaum bekannt ist – und manche glauben auch gar nicht, daß dieses Auto auch wirklich ein Sportwagen ist!'' So gab sich John R. Bond Mühe, seine Leser davon zu überzeugen. Er zitierte die von Maurice Olley niedergelegten Konstruktions-Vorgaben und nannte die Corvette einen ,,angenehmen Kontrast zu vielen Pseudo-Sportwagen, wie sie unter anderen General-Motors-Marken angeboten werden''. Dennoch: Zwischen den Zeilen standen ein paar Fragezeichen.

Als die Corvette mit V8-Motor erschien, gab sich die Motorpresse schon fröhlicher. Doch es blieb Kritik am Getriebe und an den Bremsen: ,,Davon ist der Enthusiast nicht zu überzeugen. Was soll auch der Slogan – For Experts only –? Es gehört keine besondere Begabung dazu, die Corvette zu fahren . . .'' Was man sich wünschte, war vor allem ein Getriebe mit Handschaltung. Und die hatte man endlich im 1956er Modell zu testen Gelegenheit. **Road & Track** schrieb: ,,Auf den ersten Blick meint man, drei Gänge sind nicht ausreichend. Doch die Übersetzungen sind ausgezeichnet abgestuft. 100 Meilen in zweiten Gang! Diese neue Corvette verspricht ein sehr interessanter Wagen für den Liebhaber zu werden, der ein gutes Allroundfahrzeug sucht. Es dürfte sogar Chancen in der Serienwagenklasse C (3 bis 5 Liter) haben und dort für eine gewisse Belebung sorgen . . .''

Als ,,gut bis ausgezeichnet'' beurteilten Journalisten von **Sports Car Illustrated** die 1956er Corvette. Sie stellten vor allem die Beschleunigung heraus, gemessen mit 7,4 Sekunden vom Stand auf 100 km/h. Als Spitzentempo wurden 130 Meilen = 209 km/h gemessen. Allerdings stellte man auch einige Vibrationen bei hohem Tempo fest, was auch **Road & Track** konstatierte: ,,Schütteler-scheinungen bei mehr als 100 Meilen gab es früher nicht, auch ist die Maschine, wie es scheint, etwas laut geworden . . .''

Die 1958er Corvette wurde wieder mit viel Euphorie begrüßt. ,,Zwei Vierfachvergaser oder Benzineinspritzung bei einer Kompression von 10,5 : 1! Ein Drehzahlmesser

Nr. 867 — Chevrolet Corvette Sport-Cabriolet

| | | |
|---|---|---|
| Ausrüstung | Türen: 2 | |
| | Synchromesh-Getriebe | |
| | Scheibenwaschanlage | |
| | Heizung und Entfroster | |
| | Handbrems-Warnlicht | |
| | Uhr | |
| | Elektrisch betätigtes Verdeck | |
| | Abnehmbares Kunststoff-Hardtop | |
| Motor | Steuer-PS | 23,62 |
| | Brems-PS bei U/Min. | 233/4800 |
| | Anzahl Zylinder | 8 |
| | Zylinderinhalt cm³ | 4637 |
| | Bohrung mm | 98,42 |
| | Hub mm | 76,20 |
| | Kompressionsverhältnis | 9,5:1 |
| | Verhältnis Brems-PS zu Gewicht kg/PS | 6,1 |
| Getriebe | Synchromesh-Getriebeuntersetzung | I: 2,2:1 |
| | | II: 1,66:1 |
| | | III: 1,31:1 |
| | | IV: 1:1 |
| | | R: 2,26:1 |
| Abmessungen und Gewichte | Hinterachse | 3,7:1 |
| | Länge mm | 4488 |
| | Breite mm | 1788 |
| | Höhe mm | 1320 |
| | Leergewicht fahrbereit kg | 1430 |
| | Radstand mm | 2591 |
| | Spur vorn mm | 1448 |
| | Spur hinten mm | 1499 |
| | Spurkreisdurchmesser m | 11,10 |
| | Pneugrösse | 7,3×1 |
| | Benzintankinhalt l | 62 |

bis 8000 Touren! Eine kräftiger dimensionierte Kupplung und ein selbsthemmendes Differential! Dreiganggetriebe oder Viergangautomatik zur Wahl! Sportfahrer wie auch Komfort-Liebhaber kommen voll auf ihre Kosten . . ."

Die häufigen stilistischen Änderungen, in den USA gang und gäbe, stießen nie auf Kritik. Das Cockpit, so befand die Presse, gerate von Jahr zu Jahr besser, nur sei es (1959) an der Zeit, der Corvette Sicherheitsgurte zu geben „für hartes Fahren" – obligatorisch waren sie noch nicht. Ansonsten gab es nur zwei Punkte zu kritisieren: eine gewisse Übersteuerungstendenz und ein zu langsames Ansprechen des Motors beim Druck aufs Gaspedal, was bei Schaltvorgängen zu unerwünschten Beschleunigungspausen führte.

Hin und wieder wurde auch die Kunststoff-Karosserie kritisch erwähnt, doch man konnte ihr keine Mängel nachsagen. „Sie ist tadellos verarbeitet, die Übergänge einzelner Teile sind perfekt. Man merkt, daß sich Chevrolet bei der Karosserie sehr viel Mühe gibt." Großes Lob bekamen die Sitze: „Sehr gut ausgeformt und komfortabel. Auch bei längeren Reisen keine Ermüdung am Steuer. Ein Tag in der Corvette ist kein Streß!" (1961).

Als der Sting Ray seine Aufwartung machte, wurde er mit Applaus begrüßt. Sogar die **New York Times** widmete ihm ausführlich Raum: „Wirklich gelungen! Innerhalb von zehn Jahren hat sich Amerikas einziger Sport-

*Einer der wenigen Prospekte in deutscher Sprache (für die Schweiz) beschränkte sich auf die Angabe technischer Daten und die Aufzählung der General-Motors-Händler.*

Chevrolet-Vertretungen in der Schweiz
Distributeurs Chevrolet en Suisse

| | |
|---|---|
| Aadorf | Eugen Ruckstuhl, Garage, Tel. (052) 4 73 19 |
| Aarau | F. Glaus & Co., Hohlgass-Garage, Tel. (064) 2 13 33 |
| Baden | B. Leoni AG, Bruggerstrasse 50, Tel. (056) 2 38 38 |
| Basel | Agence Américaine Automobiles S.A., Viaduktstr. 45 (Postfach 105), Tel. (061) 24 66 66 |
| Bern | Bellevue-Garage AG, Kochergasse 1, Tel. (031) 3 77 77 |
| Bern | Gebr. Bärtschi AG, Länggass-Garage, Tel. (031) 3 36 33 |
| Bern | Spiegl & Waber GmbH, Nordring 8, Tel. (031) 9 44 44 |
| Bern | Hans Lack, Ostring-Garage, Freudenbergerplatz Tel. (031) 4 49 11 |
| Biel | Burkhalter & Brändli, Garage, Freiestrasse 7, Tel. (032) 2 25 24 |
| Buchs/SG | A. Sulser, Garage, Tel. (085) 6 14 14 |
| La Chaux-de-Fonds | Garage Guttmann S.A., 110, rue de la Serre, tél. (039) 3 46 81 |
| Chur | L. Dosch, Grand-Garage, Tel. (081) 2 13 13 |
| Fribourg | L. & M. Baudère, Garage de Pérolles, tél. (037) 2 38 88 |
| Genève | Etabl. Fleury & Cie S.A., rue de la Mairie 35, tél. (022) 36 62 30 |
| Glarus | F. Schielly-Ryffel, Central-Garage, Tel. (058) 5 18 34 |
| Interlaken | E. Zimmermann, Bahnhof-Garage, Tel. (036) 2 13 15 |
| Langenthal | M. & E. Würgler, Garage, Tel. (063) 2 20 77 |
| Lausanne | Etabl. Ch. Ramuz S.A., avenue d'Echallens 2a, tél. (021) 24 04 44 |
| Liestal | Blank AG, Tiergartenweg 1, Tel. (061) 84 31 11 |
| Lugano | Eredi di N. Crescionini, Via Stefano Franscini 8, tel. (091) 2 83 43 |
| Luzern | Auto-Koch AG, Löwenplatz, Tel. (041) 2 77 77 |
| Münsingen | Ernst Marti, Garage, Tel. (031) 68 15 15 |
| Neuchâtel | Garage Schenker, Faubourg du Lac 29, tél. (038) 5 28 64 |
| Porrentruy | Périat & Pétignat, Garage des Ponts, tél (066) 6 12 06 |
| Reiden | Turmgarage, Luzernerstrasse, Tel. (062) 9 35 44 |
| Sarnen | Cl. Sidler, Garage, Brünigstrasse, Tel. (041) 85 18 88 |
| Schaffhausen/Neuhausen | Georg Neck, Garage Enge, Tel. (053) 6 94 55 |
| Schattdorf | A. Brand-Stadler, Autogarage, Tel. (044) 2 13 88 |
| Sion | G. Revaz, Garage de l'Ouest, tél. (027) 2 22 62 |
| Solothurn | E. Kupferschmid & O. Müller, Tel. (065) 2 14 42 |
| St. Gallen | E. Wagner, Centralgarage AG, Tel. (071) 22 55 22 |
| St. Maurice | Jean-Jacques Casanova, Garage, tél. (025) 3 63 90 |
| St. Moritz | Gebr. Cattaneo, Kulm-Garage, Tel. (082) 3 33 33 |
| Sursee | O. & R. Wyder, Automobile, Bahnhofstrasse, Tel. (045) 4 22 22 |
| Thun | Paul Wenger, Oberland-Garage, Tel. (033) 2 30 29 |
| Weinfelden | J. Ammann-Grünert, Central-Garage, Tel. (072) 5 13 33 |
| Wetzikon | B. Krähenmann, Werkgarage, Usterstrasse 46, Tel. (051) 97 87 02 |
| Wil | Fürstenland-Garage AG, Tel. (073) 6 04 80 |
| Winterthur | Eulach-Garage AG, Technikumstrasse 67, Tel. (052) 2 23 33 |
| Yverdon | W. Humberset, Garage des Remparts, tél. (024) 2 35 35 |
| Zug | Jos. Iten, Autos, Tel. (042) 4 23 23 |
| Zürich | AG Vertretung Agence Américaine, Dufourstrasse 23, Tel. (051) 32 72 73 |

wagen zu einem reinrassigen Fahrzeug gemausert, für Straßen und Sport gleichermaßen ausgezeichnet geeignet."

Die ausgewogene Gewichtsverteilung und die Einzelradaufhängung wurden besonders herausgestellt. **Motor Trend** schrieb: „Ein bemerkenswerter Unterschied. Der Abstand zu europäischen Sportwagen ist aufgeholt!" **Road & Track:** „Der neue Sting Ray klebt auf der Straße . . . vorbei ist die Zeit, als die Corvette mit den Hinterrädern Tänze aufführte wie ein Apache mit heißen Füßen!"

Bald war die Corvette mit Motoren bis zu 375 PS zu haben. „Obwohl so stark motorisiert, ist der Wagen lammfromm zu führen", hieß es in einem **Road & Track**-Test 1964. Und in Verbindung mit der Viergang-Automatik sei so viel Power exzellent zu beherrschen. Der Zuspruch, dessen sich die Corvette nach zehn Jahren ihrer Existenz zu erfreuen habe, sei enorm gewachsen und mit Porsche vergleichbar. Und: „Endlich hat man die Teilung der Heckscheibe wegfallen lassen – eine wesentliche Verbesserung der Sicht nach hinten." Man beeilte sich auch hinzuzufügen, daß die Schlafaugen-Scheinwerfer kein Gag der Stylisten in Detroit seien, sondern aerodynamische Vorteile brächten.

„Es gibt im Sting Ray auch Platz fürs Gepäck – nur muß es über die Sitze verstaut werden, denn der Wagen hat keine Kofferklappe. Gut wäre es, man könnte die Heckscheibe wie beim Aston Martin öffnen" (ein Wunsch, der später bei einem Sondermodell in Erfüllung ging). „Kritik möchten wir auch an den Platzverhältnissen üben: Der Fahrer sitzt recht beengt – im Vergleich zu den großzügigen Gesamtabmessungen des Wagens." Ansonsten: Höchste Anerkennung über den qualitativen Verarbeitungsstandard.

Im August 1965 wurde ein 425-PS-Wagen der Presse zur Verfügung gestellt. „Man würde sich wünschen, daß diese Corvette im Rennsport Furore macht . . . aber leider dürfte der Wagen dafür zu viel Gewicht haben", schrieb **Road & Track.** In der Tat: Die Corvette war 400 Kilogramm schwerer als die Cobra. „Es kommt bekanntlich nicht auf die Zahl der Pferdestärken an, sondern wieviele man davon zum Nutzen auf den Boden bringt . . . Viele Sportwagen hätten mehr Pfeffer nötig gehabt, aber die Corvette gehörte eigentlich nicht zu ihnen." In den siebziger Jahren wußte die Presse nicht viel Neues über „Amerikas einzigen Sportwagen" zu berichten. Man wiederholte die Hoffnungen und Erwartungen hinsichtlich motorsportlicher Engagements, auch in der europäischen Motorpresse.

Eines kurzen Schlaglichts erfreute sich die berühmte Renn-Corvette, die 1957 von Zora Arkus-Duntov präpariert worden war, in der deutschen Motorpresse. **Auto, motor und sport** berichtete über dieses Fahrzeug („das einzige Auto, mit dem sich General Motors jemals offiziell am Rennsport beteiligte") und schrieb:

„Weil die Zeit drängte, entkleidete Duntov ein Mercedes 300 SL Coupé und kopierte den Rohrrahmen. Gerüchte zu jener Zeit besagten zwar, daß der Rahmen dem des Jaguar D-Type nachgebaut wurde, doch beweist ein damals gemachtes Foto, daß es sich doch um einen 300-SL-Rahmen handelte. Er bestand aus dreieckförmig angeordneten Rohren aus Chrom-Molybdänstahl. Vorn und hinten waren stärkere Rohre zur Aufnahme der Radaufhängungen montiert."

Die Karosserien – es gab eine aus Leichtmetall und eine aus Kunststoff – wogen je 82 Kilogramm – genausoviel wie der Rohrrahmen. „Die Corvette SS geriet sehr viel leichter und kleiner als die Serienausführung . . . das Gesamtgewicht betrug nur 900 Kilogramm."

Die ersten Probefahrten, so wird berichtet, seien von Fangio und Moss absolviert worden. Im Sebring-Rennen fuhren den Wagen John Fitch und Piero Taruffi. Doch die Corvette SS beendete das Rennen nicht – wegen einer ganzen Reihe kleinerer, aber letztlich ent-

*Berichterstattung über die 250 Meilen von Daytona in „sportauto", Januar 1976. Die Corvette des Amerikaners John Greenwood hatte 725 PS unter der Haube.*

scheidender Pannen. Der Wagen kam später ins Museum – er wurde nie mehr eingesetzt.
Privat eingesetzte Exemplare tauchten indessen an zahlreichen Rundstreckenrennen immer wieder auf, in den USA und auch anderswo – mit unterschiedlichem Erfolg. Ein hochfrisierter Sting Ray mit über 700 PS beteiligte sich zum Beispiel an den 250 Meilen von Daytona 1976 und anderen Wettbewerben. Yörn Pugmeister berichtete in **sport auto** über das Corvette-Monster:
,,Die äußere Form erinnert noch an ‚den' amerikanischen Sportwagen, aber John Greenwood schuf sich seine eigene Corvette. Dazu fräste er erst einmal das Chassis des Ursprungsautos weitgehend weg und ersetzte es durch einen Rohrrahmen; dann spendierte er dem Mobil neue Bremsen vorn und hinten . . .'' Als Motor installierte Greenwood einen 8,2 Liter mit 725 PS. Mit einem Schnitt von 188 km/h wurde Greenwood als Sieger abgewunken – mehr als eine halbe Minute vor Brian Redman auf einem BMW 3,0 CSL.
Die Corvette der frühen achtziger Jahre gab ebenfalls – noch – zu keinen Klagen Anlaß. ,,Power ist genügend vorhanden; in 7,2 Sekunden ist man auf 100 km/h'', schrieb John Lamm 1983. ,,Aber die Corvette bietet mehr als nur Tempo. Zum Beispiel einen völlig neuen Sitzkomfort. Jezt hat man nicht mehr das Gefühl, in einem riesigen Schuh zu sitzen, der dennoch zu eng ist . . .'' Es gebe eine gute Rundumsicht, eine überaus reichhaltige Instrumentierung (wahlweise auf metrische Anzeige umschaltbar, was sehr begrüßt wurde) und erfreulicherweise keine überflüssigen Mätzchen. ,,Ist die Corvette der beste Exote der Welt?'' Man verglich sie mit einem Porsche 928 S und einem Ferrari 308 GTBi und kam zu dem Schluß, daß der Amerikaner gar nicht so schlecht abschneide . . . Vor allem sei der Wagen im Design seinen europäischen Rivalen überlegen (meinte Autor Jonathan Thompson): ,,Er ist das am besten aussehende Auto, das heute auf dem Markt ist . . . sowohl aufregend als auch geschmackvoll und elegant . . . Mit dieser Corvette haben sich die Schlaglichter der Karosserie-Designbranche von Europa nach Amerika herübergewendet . . .'' Ein Kompliment, das Jerry Palmer zugedacht war, dem ,,Corvette Body Designer'', der auch für den Camaro und Chevrolet Vega verantwortlich gezeichnet hatte.

*Dreamcar der neunziger Jahre: Der 1989 in Genf gezeigte Chevrolet Corvette mit 5,7-Liter-Motor, ein Vierventiler mit allen Attributen modernen Hochleistungs-Motorenbaus.*

# Technische Daten

| Modelljahr | Motorenspezifikation | | Gemischaufbereitung | Anmerkungen |
|---|---|---|---|---|
| 1953 | 6 Zylinder Blue Flame | | 3 Flachstrom-Vergaser | 300 Vorserienwagen; alle weiß |
| 1954 | 235 c.i. (3858 ccm) | 150 PS/4200 ohv | 3 Flachstrom-Vergaser | Farbe jetzt auch rot; blau; kupfer |
| 1955 | Blue Flame 235 c.i. (3858 ccm)<br>Turbo Fire V-8 265 c.i. (4343 ccm) | 155 PS/4200 ohv<br>195 PS/4400 | 3 Flachstrom-Vergaser<br>1 Fallstrom-Vierfach-Vergaser | Neu: V8-Motor; 12-Volt-Anlage; Einige Wagen mit 3-Gang-Schaltung |
| 1956 | V-8 265 c.i. (4343 ccm) | 210 PS/5200 ohv<br>225 PS/5200 ohv | 1 Fallstrom-Vierfach-Vergaser<br>2 Fallstrom-Vierfach-Vergaser | Neue Karosserie; 3-Gang-Getriebe; Auf Wunsch Powerglide |
| 1957 | V-8 283 c.i. (4638 ccm)<br><br><br>Competition | 220 PS/4800 ohv<br>245 PS/5000 ohv<br>250 PS/5000 ohv<br>283 PS/6200 ohv | 1 Fallstrom-Vierfach-Vergaser<br>2 Fallstrom-Vierfach-Vergaser<br>Ramjet Fuel Injection<br>Ramjet Fuel Injection | Auf Wunsch Viergang-Schaltung |
| 1958 | V-8 283 c.i. (4638 ccm)<br><br><br>Competition | 230 PS/4800 ohv<br>245 PS/5000 ohv<br>250 PS/5000 ohv<br>290 PS/6200 ohv | 1 Fallstrom-Vierfach-Vergaser<br>2 Fallstrom-Vierfach-Vergaser<br>Ramjet Fuel Injection<br>Ramjet Fuel Injection | Neuer Grill und Doppelscheinwerfer, mehr Chromschmuck |
| 1959 | V-8 283 c.i. (4638 ccm) | 230 PS/4800 ohv<br>245 PS/5000 ohv<br>250 PS/5000 ohv<br>270 PS/5400 ohv<br>290 PS/6200 ohv | 1 Fallstrom-Vierfach-Vergaser<br>2 Fallstrom-Vierfach-Vergaser<br>Ramjet Fuel Injection<br>2 Fallstrom-Vierfach-Vergaser<br>Ramjet Fuel Injection | Chrom wird wieder reduziert |
| 1960 | V-8 283 c.i. (4638 ccm) | 230 PS/4800 ohv<br>245 PS/5000 ohv<br>270 PS/5400 ohv<br>275 PS/5800 ohv<br>315 PS/6400 ohv | 1 Fallstrom-Vierfach-Vergaser<br>2 Fallstrom-Vierfach-Vergaser<br>2 Fallstrom-Vierfach-Vergaser<br>Ramjet Fuel Injection<br>Ramjet Fuel Injection | Auf Wunsch elektrische Fensterheber |
| 1961 | V-8 283 c.i. (4638 ccm) | wie 1960 | wie 1960 | Neuer Grill; neues Heck |
| 1962 | V-8 327 c.i. (5359 ccm) | 250 PS/4400 ohv<br>300 PS/5000 ohv<br>340 PS/6000 ohv<br>360 PS/6000 | 1 Fallstrom-Vierfach-Vergaser<br>1 Fallstrom-Vierfach-Vergaser<br>1 Fallstrom-Vierfach-Vergaser<br>Ramjet Fuel Injection | Neuer Motor „327“ |

| Modelljahr | Motorenspezifikation | | Gemischaufbereitung | Anmerkungen |
|---|---|---|---|---|
| 1963 | V-8 327 c.i. (5359 ccm) | wie 1962 | wie 1962 | Neue Karosserie; Neu: Fastback-Coupé |
| 1964 | V-8 327 c.i. (5359 ccm) | 250 PS/4400 ohv<br>300 PS/5000 ohv<br>365 PS/6200 ohv<br>375 PS/6200 ohv | 1 Fallstrom-Vierfach-Vergaser<br>1 Fallstrom-Vierfach-Vergaser<br>1 Fallstrom-Vierfach-Vergaser<br>Fuel-Injection | Verstellbare Stoßdämpfer |
| 1965 | V-8 327 c.i. (5359 ccm)<br><br><br><br><br>V-8 396 c.i. (6489 ccm) | 250 PS/4400 ohv<br>300 PS/5000 ohv<br>350 PS/5800 ohv<br>365 PS/6200 ohv<br>375 PS/6200 ohv<br>PS/4800 ohv | 1 Fallstrom-Vierfach-Vergaser<br>1 Fallstrom-Vierfach-Vergaser<br>1 Fallstrom-Vierfach-Vergaser<br>1 Fallstrom-Vierfach-Vergaser<br>Fuel-Injection<br>2 Fallstrom-Vierfach-Vergaser | Vier Scheibenbremsen |
| 1966 | V-8 327 c.i. (5359 ccm)<br><br><br>Turbo-Jet 427 c.i. (6997 ccm)<br>V-8 | 300 PS/5000 ohv<br>350 PS/5800 ohv<br><br>390 PS/5200 ohv<br>425 PS/5600 ohv | 1 Fallstrom-Vierfach-Vergaser<br>1 Fallstrom-Vierfach-Vergaser<br><br>1 Fallstrom-Vierfach-Vergaser<br>1 Fallstrom-Vierfach-Vergaser | Neuer Turbo-Jet 427-c.i.-Motor |
| 1967 | V-8 327 c.i. (5359 ccm)<br><br><br>Turbo-Jet 427 c.i. (6997 ccm)<br>V-8 | 300 PS/5000 ohv<br>350 PS/5800 ohv<br><br>390 PS/5400 ohv<br>400 PS/5400 ohv<br>435 PS/5800 ohv | 1 Fallstrom-Vierfach-Vergaser<br>1 Fallstrom-Vierfach-Vergaser<br><br>1 Fallstrom-Vierfach-Vergaser<br>3 Doppelvergaser<br>3 Doppelvergaser | Neue Doppelvergaser-Hochleistungsmotoren |
| 1968 | V-8 327 c.i. (5359 ccm)<br>V-8 396 c.i. (6489 ccm)<br>V-8 427 c.i. (6997 ccm) | 300 PS/5000 ohv<br>350 PS/5200 ohv<br>390 PS/5400 ohv<br>400 PS/5400 ohv<br>435 PS/5800 ohv | 1 Fallstrom-Vierfach-Vergaser<br>1 Fallstrom-Vierfach-Vergaser<br>1 Fallstrom-Vierfach-Vergaser<br>3 Doppelvergaser<br>3 Doppelvergaser | Neues Karosserie-Styling |
| 1969 | V-8 350 c.i. (5735 ccm)<br><br><br>V-8 427 c.i. (6997 ccm) | 300 PS/4800 ohv<br>350 PS/5600 ohv<br><br>390 PS/5400 ohv<br>400 PS/5400 ohv<br>425 PS/5600 ohv<br>430 PS/5200 ohv<br>435 PS/5800 ohv | 1 Fallstrom-Vierfach-Vergaser<br>1 Fallstrom-Vierfach-Vergaser<br><br>1 Fallstrom-Vierfach-Vergaser<br>1 Fallstrom-Vierfach-Vergaser<br>1 Fallstrom-Vierfach-Vergaser<br>1 Fallstrom-Vierfach-Vergaser<br>3 Doppelvergaser | Neuer 350-c.i.-Motor |

## Technische Daten

| Modelljahr | Motorenspezifikation | | Gemischaufbereitung | Anmerkungen |
|---|---|---|---|---|
| 1970 | V-8 350 c.i. (5735 ccm)<br><br><br>V-8 454 c.i. (7440 ccm) | 300 PS/4800 ohv<br>350 PS/5600 ohv<br>370 PS/6000 ohv<br>390 PS/4800 ohv<br>460 PS/5600 ohv | 1 Fallstrom-Vierfach-Vergaser<br>1 Fallstrom-Vierfach-Vergaser<br>1 Fallstrom-Vierfach-Vergaser<br>1 Fallstrom-Vierfach-Vergaser<br>1 Fallstrom-Vierfach-Vergaser | Neuer 454-c.i.-Motor |
| 1971 | V-8 350 c.i. (5735 ccm)<br><br>V-8 454 c.i. (7440 ccm) | 270 PS/4800 ohv<br>330 PS/5600 ohv<br>365 PS/4800 ohv<br>425 PS/5600 ohv | 1 Fallstrom-Vierfach-Vergaser<br>1 Fallstrom-Vierfach-Vergaser<br>1 Fallstrom-Vierfach-Vergaser<br>1 Fallstrom-Vierfach-Vergaser | |
| 1972 | V-8 350 c.i. (5735 ccm)<br><br>V-8 454 c.i. (7440 ccm) | 200 PS/4400 ohv<br>255 PS/5600 ohv<br>270 PS/4000 ohv | 1 Fallstrom-Vierfach-Vergaser<br>1 Fallstrom-Vierfach-Vergaser<br>1 Fallstrom-Vierfach-Vergaser | Drastische Leistungsminderungen durch neue Abgasgesetze |
| 1973 | V-8 350 c.i. (5735 ccm)<br><br>V-8 454 c.i. (7440 ccm) | 190 PS/4400 ohv<br>250 PS/5200 ohv<br>275 PS/4000 ohv | 1 Fallstrom-Vierfach-Vergaser<br>1 Fallstrom-Vierfach-Vergaser<br>1 Fallstrom-Vierfach-Vergaser | |
| 1974 | V-8 350 c.i. (5735 ccm)<br><br>V-8 454 c.i. (7440 ccm) | 195 PS/4400 ohv<br>250 PS/5200 ohv<br>270 PS/4400 ohv | 1 Fallstrom-Vierfach-Vergaser<br>1 Fallstrom-Vierfach-Vergaser<br>1 Fallstrom-Vierfach-Vergaser | Neues aufprallsicheres Heck |
| 1975 | V-8 350 c.i. (5735 ccm)<br><br>V-8 454 c.i. (7440 ccm) | 198 PS/4400 ohv<br>253 PS/5200 ohv<br>247 PS/4400 ohv | 1 Fallstrom-Vierfach-Vergaser<br>1 Fallstrom-Vierfach-Vergaser<br>1 Fallstrom-Vierfach-Vergaser | |
| 1976 | V-8 350 c.i. (5735 ccm) | 167 PS/3800 ohv<br>208 PS/4800 ohv | 1 Fallstrom-Vierfach-Vergaser<br>1 Fallstrom-Vierfach-Vergaser | |
| 1977 | V-8 350 c.i. (5735 ccm) | 183 PS/4000 ohv<br>213 PS/5200 ohv | 1 Fallstrom-Vierfach-Vergaser<br>1 Fallstrom-Vierfach-Vergaser | |
| 1978 | V-8 350 c.i. (5735 ccm) | 180 PS/4000 ohv<br>210 PS/5200 ohv | 1 Fallstrom-Vierfach-Vergaser<br>1 Fallstrom-Vierfach-Vergaser | Scheibenbremsen serienmäßig |
| 1979 | V-8 350 c.i. (5735 ccm) | 185 PS/4000 ohv<br>220 PS/5200 ohv | 1 Fallstrom-Vierfach-Vergaser<br>1 Fallstrom-Vierfach-Vergaser | Einzelradaufhängung hinten |

# Technische Daten

| Modell-jahr | Motoren-spezifikation | | Gemischaufbereitung | Anmerkungen |
|---|---|---|---|---|
| 1980 | V-8 350 c.i. (5735 ccm) | 195 PS/4000 ohv<br>225 PS/5200 ohv | 1 Fallstrom-Vierfach-Vergaser<br>1 Fallstrom-Vierfach-Vergaser | Gewichtsreduzierung um 140 kg |
| 1981 | V-8 305 c.i. (5001 ccm)<br>V-8 350 c.i. (5735 ccm) | 186 PS/4200 ohv<br>197 PS/4400 ohv<br>238 PS/5200 ohv | 1 Fallstrom-Vierfach-Vergaser<br>1 Fallstrom-Vierfach-Vergaser<br>1 Fallstrom-Vierfach-Vergaser | |
| 1982 | V-8 350 c.i. (5735 ccm) | 197 PS/4400 ohv | 1 Fallstrom-Vierfach-Vergaser | Collector edition:<br>Aufklappbare Heckscheibe |
| 1983 | V-8 350 c.i. (5735 ccm) | 204 PS/4200 ohv | Fuel Injection | Völlig neues Modell |
| 1984/85 | V-8 350 c.i. (5735 ccm) | 208 PS/4200 ohv | Fuel Injection | |
| 1986 | V-8 350 c.i. (5735 ccm) | 208 PS/4200 ohv | Fuel Injection | Neues Cabriolet |

## Technische Daten

Schrader-Motor-Chronik

# Amerikanische Station Wagons und Woodies

## 1938 – 1960

Eine Dokumentation von Halwart Schrader

# VORWORT

Kombiwagen: Nichts besonderes. Oder doch? Die Exoten unter ihnen sind durchaus Sammlerstücke, wie zum Beispiel die amerikanischen Station Wagons, vor allem jene mit einer Karosserie in Holz-Stahl-Gemischtbauweise, »Woodies« genannt. Ihnen ist dieser Band der Reihe Schrader-Motor-Chronik gewidmet.

Lassen Sie sich also durch einen nach dem Markenalphabet geordneten Querschnitt der US-Straßenkreuzer-Geschichte führen. Er zeigt Fahrzeuge, die wir heute bei uns kaum mehr zu sehen bekommen, und auch zu »ihrer Zeit« waren sie stets die Ausnahme im Straßenbild. Gleichwohl gibt es in fast allen Prospekten der amerikanischen Hersteller – und ihrer europäischen Repräsentanten – sehr ausführliche Beschreibungen der Kombiwagen, und in der oft maßlosen zeichnerischen Übertreibung ihrer Dimensionen vermitteln sie ein deutliches Bild des Werbestils, der in den fünfziger Jahren in den USA vorherrschte.

Weiter hinten im Buch werden zum Vergleich einige europäische Kombiwagen, zum Teil ebenfalls im Woody-Stil, gezeigt. Die Auswahl erfolgte eher individuell, wie auch die der amerikanischen Wagen – es würde ins Unendliche führen, wollte man eine komplette Übersicht liefern. Was auch nicht das Anliegen dieser Reihe ist. Einen umfassenden Eindruck über die Automobil-Architektur jener Zeit, dessen bin ich mir sicher, vermittelt die Dokumentation dennoch.

Halwart Schrader
Herausgeber

Schon in den frühen zwanziger Jahren gab es vereinzelt Automobile, die sich durch eine Besonderheit von normalen Serien-Personenwagen unterschieden: Sie besaßen eine zusätzliche Tür im Heck und konnten durch die Herausnahme der hinteren Sitzbank in einen Lieferwagen verwandelt werden. Meist wurden solche Aufbauten nach den Wünschen des Kunden in einer Zeit angefertigt, als es noch üblich war, eine individuell angefertigte Karosserie auf ein separates Fahrgestell zu setzen.

Kombinations-Kraftwagen, wie sie in Deutschland genannt wurden, gab es erstmals bei Ford auf Basis des T-Modelles in den zwanziger Jahren. Aber diese Spezialanfertigungen wurden nicht in den Ford-Werken gebaut, sondern von Zulieferfirmen, die für Ford-Automobile eine Fülle von Sonderaufbauten anboten. In Europa sah man serienmäßig hergestellte Kombinations-Kraftwagen erst gegen Ende der dreißiger Jahre; zum Beispiel gab es ein solches Modell bei Wanderer.

In den USA hatte der Mehrzweck-Personenwagen nach Vorbild des Ford sehr schnell seinen Markt gefunden. Viele Hersteller boten diese praktische Kombinations-Limousine an, die sich mit wenigen Handgriffen in ein Nutzfahrzeug umrüsten ließ, ohne ihren Pkw-Charakter zu verlieren. Begonnen hatte es mit einer etwas größer dimensionierten Kofferklappe, die ein Beladen des Wagens mit größereren Gepäckstücken erlaubte; aus dieser Klappe wurde eine große Tür, die schließlich auch ein Rückfenster aufwies. Einen richtigen Kombi, in Amerika Station Wagon genannt, boten nach Ford 1929 die Firma Fargo, dann ab 1937 Dodge, ein Jahr später auch Plymouth und Pontiac an. Das Wort »Station« hat in den USA viele Bedeutungen; man verwendete es für einen Handelsposten im Fernen Westen einst ebenso wie für einen

*Ford V8 von 1938 als Station Wagon mit dem typischen Holz/Stahl-Aufbau. Ein Sammlerstück von hohem Reiz.*

*Buick Roadmaster Estate Wagon 1950 – erstmals mit einer Windschutzscheibe, die keine senkrechte Strebe mehr aufweist. Der Herr im hellen Zweireiher demonstriert Besitzerstolz.*

Bahnhof, für ein Postamt wie für eine Tankstelle, für ein Farmhaus wie für ein Militärdepot.
Das Auto als sogenannter Woody (wood = Holz), heute ein begehrtes Sammlerstück, ist keine Erfindung der Nachkriegszeit. Schon lange vorher stellte man – wiederum vorzugsweise in den USA – Aufbauten her, deren Holzgerüst, meist aus Esche, mit Holztäfelung ausgekleidet war. Der

relativ hohe handwerkliche Aufwand bei der Herstellung dieser Karosserien wog die Einsparung des Stahlblechanteils auf, den eine »all steel«-Karosserie aufwies. Die Aufbauten waren zudem leichter, was eine höhere Nutzlast bedeutete (oder dem Hersteller die Möglichkeit gab, dafür noch ein paar Kilo Chromzierat draufzupacken). Aber die Aufbauten neigten auch dazu, sich zu verziehen, denn Holz pflegt zu »arbeiten«. Schlecht sitzende Türen waren das mindeste, was man im Laufe der Zeit bei einem echten Woody zu gewärtigen hatte.

Es wurde schick, einen solchen Wagen zu besitzen, und die Automobilhersteller gaben sich viel Mühe, das Styling dieser Fahrzeuge so attraktiv wie möglich zu machen. 1940 hatten fast alle Personenwagen-Hersteller Amerikas einen Woody-Kombi im Programm, ausgenommen Luxusfabrikate wie Cadillac. Doch Buick, Packard, Chrysler, Pontiac und so weiter – sie alle boten Woodies an.

In der frühen Nachkriegszeit bis in die fünfziger Jahre hinein waren die Woodies ebenfalls wieder da. Vor allem auch in Gestalt von zwei- und viertürigen Personenwagen; Chrysler bot sogar ein bildhübsches Cabriolet an.

Inzwischen sprach ein weiterer Grund für diese rustikale Bauweise: Die Verknappung von Karosseriestahl. Der Zweite Weltkrieg und alsbald auch der Koreakonflikt hatten in den USA die Stahlpreise stark ansteigen lassen, so daß es durchaus Sinn machte, Woodies im Programm zu behalten.

Der Kombiwagenanteil am Gesamtprogramm amerikanischer Personenwagen nahm ständig zu. Es gab die raffiniertesten Ausführungen mit zweigeteilten Hecktüren, einer dritten Sitzbank oder Mechanismen, durch die man die Bänke so umlegen konnte, daß sie eine ebene Bodenfläche bildeten.

Der Station Wagen wurde zu einem Kultfahrzeug. Und als die Holz-Ära vorüber war, behielten etliche Hersteller die Woody-Mode bei, indem sie die Flanken ihrer nunmehr wieder gänzlich aus Stahl hergestellten Aufbauten mit Folien beklebten, die Sperrholz-Maserung vortäuschten. Oder man übernahm einfach nur die Ausformung des Fachwerks, um sie durch eine zweifarbige Lackierung noch zu betonen.

Der Woody-Stil kam schon sehr bald nach Europa herüber. Teils aus modischen Motiven, teils auch aus Gründen der bei uns ebenfalls spürbaren Stahlverteuerung gab es Holzaufbauten in kleineren oder größeren Serien bei Fiat (500 Giardiniere, Belvedere), DKW (Auto Union Universal Kombiwagen), Goliath, Tempo, Morris, Austin, Ford GB, Alfa Romeo und anderen. In Einzelanfertigung entstanden solche Aufbauten auch auf Fahrgestellen von Rolls-Royce, Lea Francis, Alvis. Die Engländer nannten diese Autos – wie alle Kombiwagen – »Estates« (estate = Gutshof).

Auch in Europa erhielt der Kombiwagen einen zunehmenden Stellenwert. Der Caravan von Opel oder der Turnier von Ford, die Commerciale- und Familiale-Versionen französischer Fabrikate fanden rasch ihren Markt. Hersteller von Heckmotor-Fahrzeugen taten sich dabei etwas schwerer, ihre Autos als Kombiwagen herauszubringen; bei VW dauerte es beispielsweise bis zum Erscheinen des Typ 3 (1500), ehe sich ein solches Konzept realisieren ließ. Weiter hinten in diesem Buch werden ein paar klassische Vertreter der europäischen Kombiwagen-Prominenz aus den fünfziger Jahren mit aufgeführt.

Der amerikanische Station Wagon begann gegen Ende der fünfziger Jahre ein Eigenleben zu führen. Anfangs nur ein Derivat des Serien-Personenwagens, dem man die Verwandtschaft zum Pkw der gleichen Baureihe sofort ansah, verselbständigten sich allmählich die Konstruktionen; es entstanden die ersten Großraumkombis wie der Corvair Greenbrier. In Europa hießen die Gegenstücke VW Typ 2 (»Bully«), Lloyd LT 400 oder Fiat Multipla. Auch ihre Konstruktionen entstammten dem Personenwagenbau, mutierten aber zu eigenen Baureihen.

Sonderfahrzeuge in Kombiwagen-Bauweise wie Ambulanz- oder Bestattungswagen blieben hier unberücksichtigt; sie gab es auf Personenwagen-Chassis seit jeher. Ältere Modelle der in Serie gebauten Station Wagons made in USA, vor allem in Woody-Ausführung, genießen in jedem Fall einen hohen Liebhaberstatus, keineswegs nur in ihrem Heimatland.

# Buick

Buick: Eine der Traditionsmarken (gegründet 1903) im General-Motors-Konzern, und schon seit 1940 mit einem Woody-Kombiwagen mit von der Partie. Bei Buick sprach man übrigens nicht von einem Station Wagon, sondern – nach der feinen englischen Art – von einem »Estate Wagon«.

1940 kostete der Buick-Kombi 1242 Dollar, (Weißwandreifen erhielt man für 13 Dollar Aufpreis), und 1949 stand er mit knapp 2500 Dollar im Katalog. Buick Kombis der vierziger und fünfziger Jahre sind stilistisch besonders reizvoll und für ihre Ära typisch.

*Rechts: Der Buick Roadmaster Estate Wagon von 1957 hatte 3239 mm Radstand und einen 300-PS-V8-Motor unter der Haube. Er war ein schwerer Brocken und konnte wie ein Lastwagen beladen werden.*

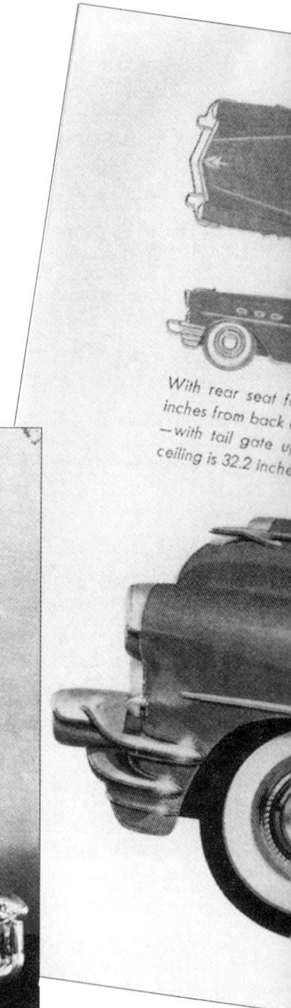


*Unten: Jedes Jahr gab es bei Buick – wie bei den meisten amerikanischen Personenwagen – einen anderen Kühlergrill. Dieser Woody ist von 1949. Die Baureihe »Super« hatte drei Bullaugen an der Seite, beim »Roadmaster« waren es vier.*

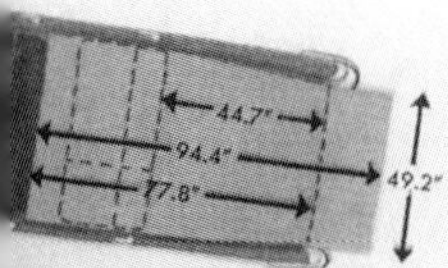

load space here is 94.4
o tip of lowered tail gate
es. Height from floor to

In the 1956 CENTURY edition of the Buick Estate Wagon, custom quality goes all the way. You get a big all-steel body so roomy it can handle practically any assignment you give it —so beautiful it turns heads wherever you go. You get 255 horsepower on a nimble 122-inch wheelbase. You get a custom-finished interior in your choice of five colors—and fully carpeted floors front and rear. And for a modest extra price, you get Buick's convenient 1/3 and 2/3 split rear seat so you can carry passengers and cargo at the same time.

***Additional* CENTURY *Features for 1956 included a standard equipment:*** *Variable Pitch Dynaflow, 4-Barre Airpower Carburetor, foam rubber seat cushions, automatic trunk light, electric clock, rear license plate frame and the new sealed beam headlights. In Convertible, electrically operated windows and horizontal front seat adjustment, plus outside rear-view mirrors.*

CENTURY 6-PASSENGER 4-DOOR ESTATE WAGON, MODEL 69, 122-IN. WHEELBASE, 255 HORSEPOWER

These two SPECIALS carry the biggest story of Buick's special price news for 1956. Above is Buick's lowest-priced car—the 2-Door Sedan. It's sweep-ahead styled, 6-passenger big, with 220 of Buick's finest horsepower.

Below is Buick's swift-lined beauty for work and play—the all-steel Estate Wagon—thrillingly beautiful—mightily powered—wonderfully handy—*and priced for the thrifty.*

So whether your tastes run to the design of a Sedan or versatility of an Estate Wagon—you have here two good budget reasons why 1956's beacon light in low-priced cars is the Buick SPECIAL Series.

The new SPECIAL Estate Wagon puts spacious
ness on a par with beauty. Rear compartme
is 94.4 inches deep from back of front seat
lowered tail gate edge. Height 32.2 inche
79.8 cubic feet load space in all.

SPECIAL 6-PASSENGER 4-DOOR ESTATE WAGON, MODEL 49, 122-IN. WHEELBASE, 220 HORSEPOWER

*Buick Special Estate Wagon, zwei- oder auch dreifarbig lackiert, Modell 1956.*

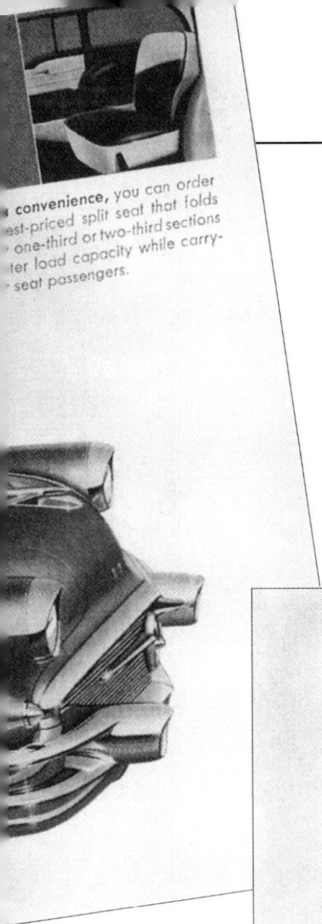


Links: *Aus einem 1956er Buick-Prospekt, der in einem extrem breiten Querformat gehalten war. »Dieser Wagen trägt unser Banner voran!« lautete der Slogan, und: »Hier beginnt ein neues, großartiges Abenteuer des Autoreisens...«*

*Unten: Buick Century Estate Wagon. Bei ausgeklappter Hecktür hatte man eine Ladefläche von 1,98 m. Der Radstand des Wagens betrug 3099 mm. Innen wie außen war der Wagen zweifarbig gehalten.*

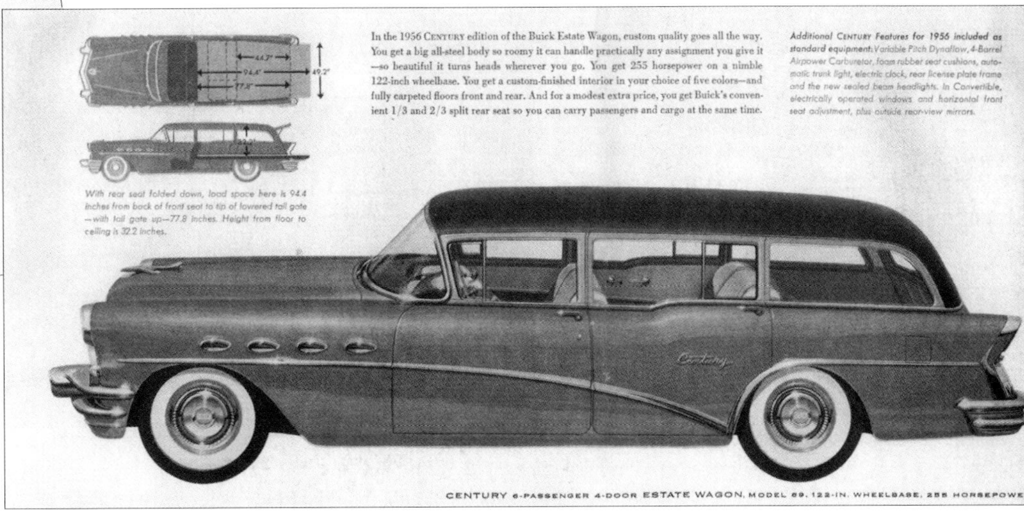

**Here, on this page,** you are face to face with the rest of the sleek and glamorous B-58 Buicks.

To the right you see Buick's sleek, glamorous and versatile Estate Wagons in 3 price ranges.

Just below is the SUPER Series—and you will find these are cars with a great new spaciousness and spirit in the medium-priced field.

At lower left is the CENTURY Series—and for 1958 it adds new sparkle to its name as America's most spectacular performance buy.

Then over at the lower right, you see the brilliant SPECIAL Series, and only the price tags will tell you that these Buicks are the thriftiest buys in the family.

When you've finished studying these stunning new B-58 Buicks, come take one out on the road. Try the wondrous luxury of the new Buick ride—the full authority of its new performance—its joyous ease of handling. See for yourself that here is truly the first big car that's light on its feet. Do that today.

# BUICK *Estat*

**CENTURY** 4-Doo

*Ende 1957 erschien dieser Prospekt, die 1958er Modelle ankündigend. Den Radstand von 3099 mm hatten alle Fahrzeuge gemeinsam, doch der Century Caballero wartete mit 300 PS auf (SAE). Schwerer Chromschmuck zeichnete diesen General-Motors-Jahrgang aus.*

*Gar nicht einmal sehr übertrieben sind diese farbigen Abbildungen (es sind keine Fotos!). Da die Fahrzeuge keine Einzelsitze, sondern durchgehende Sitzbänke aufwiesen, die jeweils drei Personen Platz boten, sprach man guten Gewissens von Sechssitzern.*

*Oben: Chevrolet 3106 Suburban 1951; links: 210 Handiman Station 1953; rechts: Chevi Station 1941.*

Kombiwagen waren bei Chevrolet, ebenfalls zu General Motors gehörend, seit 1947 im Programm. In den 50er und 60er Jahren spielten diese Modelle bei Chevrolet eine wichtige Rolle.

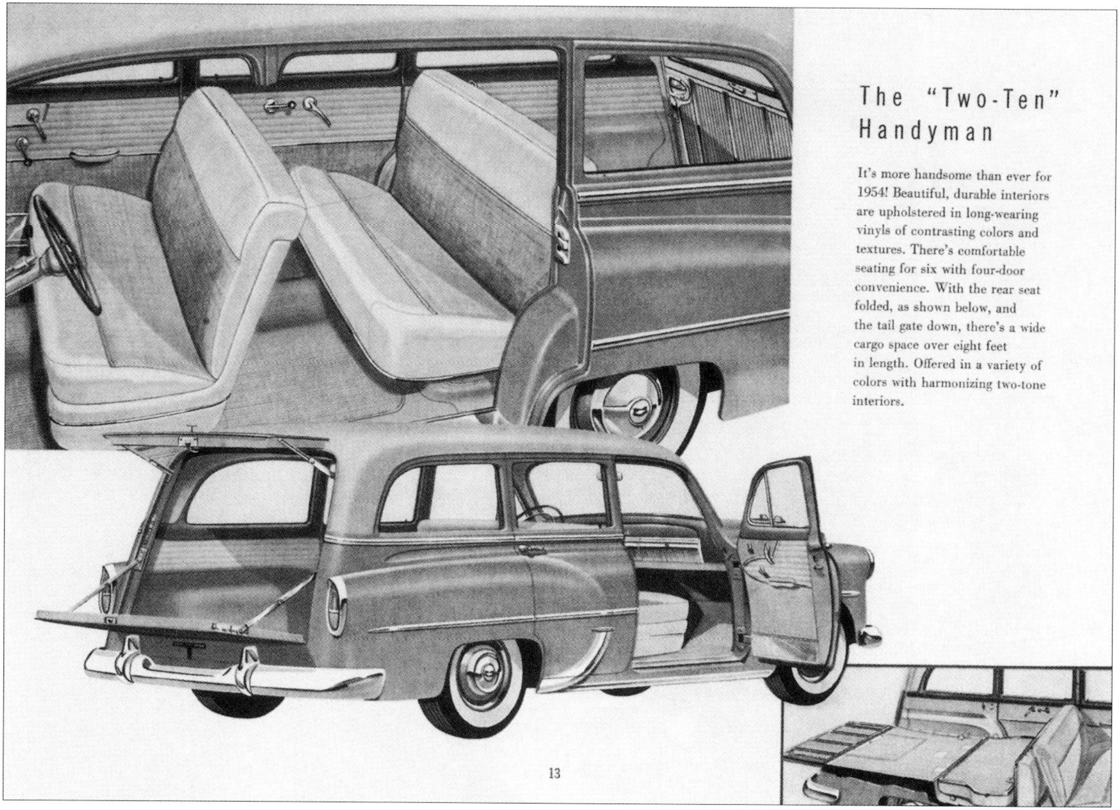


*Chevrolet Handiman 1954, Baumuster 210 (»Two-Ten«). Die in der Waagerechten zweigeteilte Heckklappe hatte sich bei amerikanischen Kombiwagen überall durchgesetzt.*

## The "One-Fifty" Handyman

Outstanding beauty and utility are yours in Chevrolet's lowest priced station wagon. The six-passenger, all-steel body has four doors for extra convenience. And the rear seat folds flat to give you a wide, level floor for extra cargo.

## The Bel Air Townsman

This beautiful new addition to the Chevrolet Bel Air series accommodates eight passengers in roomy comfort—with plenty of load space besides. The two rear seats can easily be removed for additional cargo capacity. All steel body with beautiful simulated wood grain trim, color-keyed interior.

12

*Auch den berühmten Chevrolet Bel Air gab es in einer Station-Wagon-Ausführung. Unterhalb der Fenster eine Andeutung von Holzpaneelen – aber die waren ein Imitat.*

*1955er Chevrolet Nomad, ein zweitüriger Kombi. Im Unterschied zu den Modellen Townsman Wagon V8 und Bel Air Beauville bekam man den Nomad auch mit einem Sechszylindermotor.*

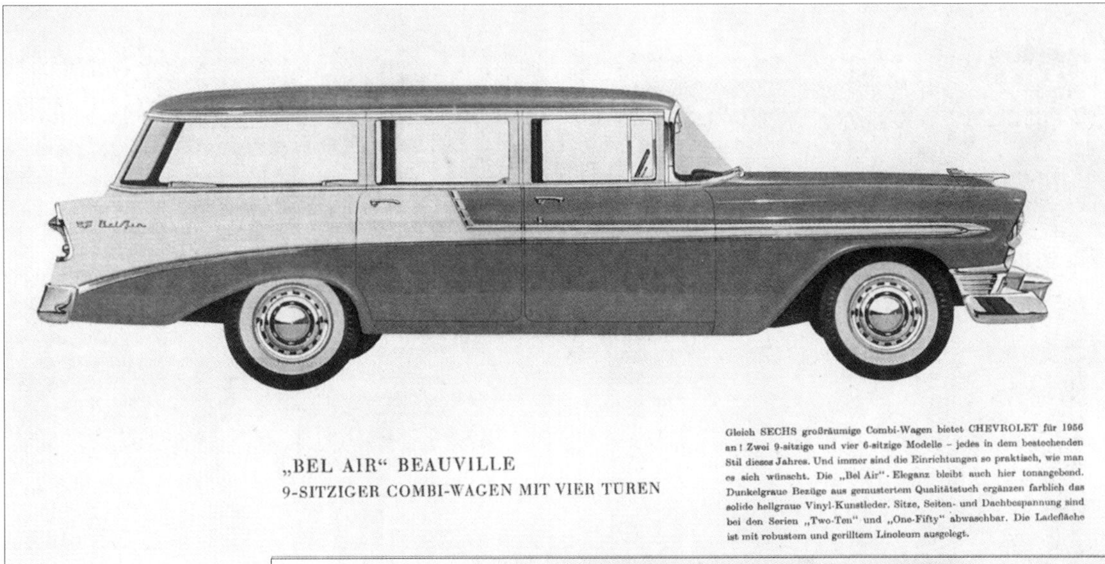

*Oben: Chevrolet Bel Air Beauville, einer der besseren Chevi-Kombis, selbstverständlich in zweifarbigem Finish.*

*Rechts: Bel Air Nomad 1956. Es gab Versionen mit 140, 162, 170, 205 oder 225 PS. Die Chromleisten auf der unteren Heckklappe waren reines Dekor.*

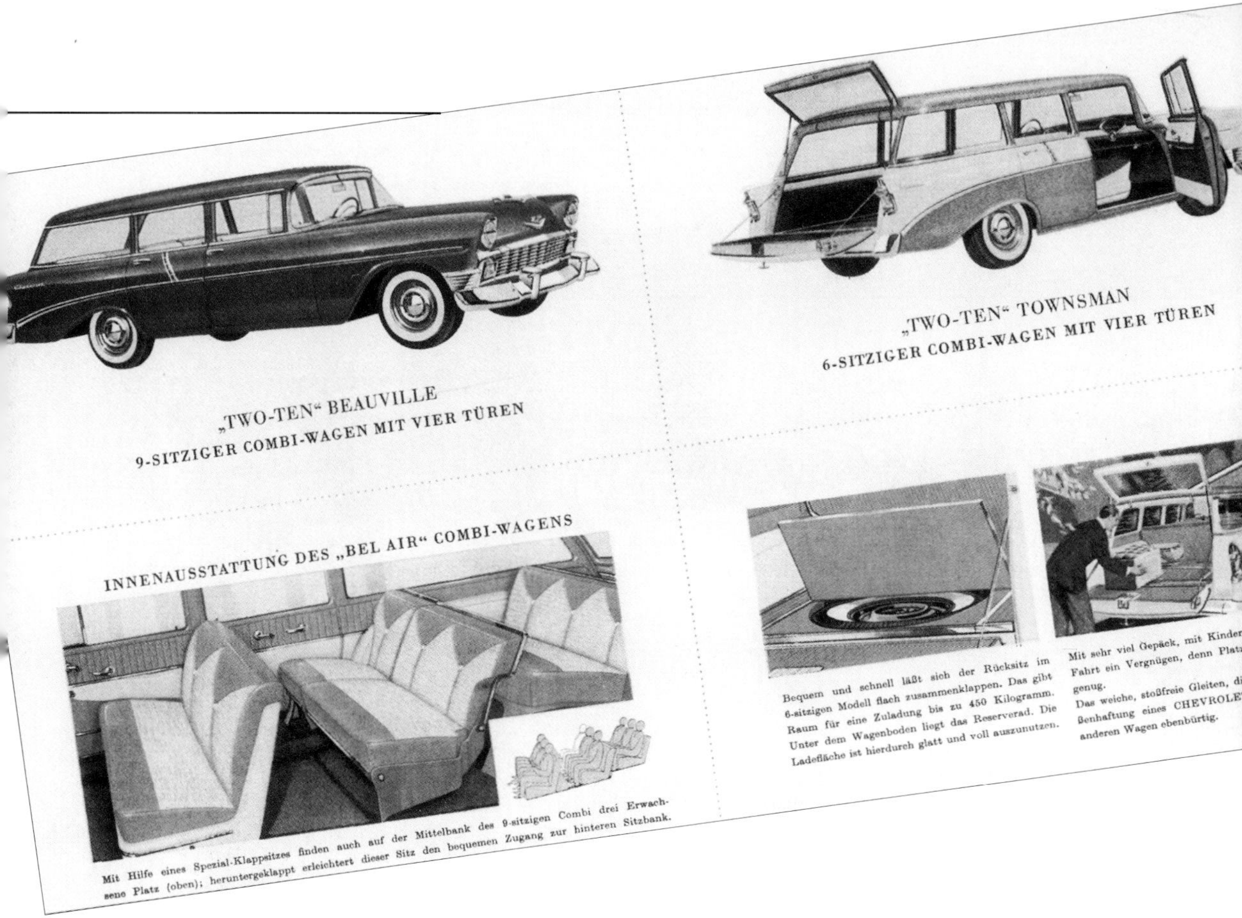


*Aus einem für die Schweiz gedruckten Chevrolet-Prospekt von Ende 1955, der die Modellplatte 1956 vorstellt. Der Motor, genannt »Super Turbo-Fire V8«, wurde als ein Aggregat der Spitzenklasse bezeichnet: »Man erkennt ihn an den zwei Auspuffrohren!«. Der Sechszylinder trug den Namen »Blue Flame 140«.*

*Chevrolet 150 (»One-Fifty«) Modelljahrgang 1957. Einen Sechssitzer dieser Größenordnung mit nur zwei Türen anzubieten, war damals nicht ungewöhnlich, viele solcher Autos fanden ihre Abnehmer.*

*Unten: 1959er Chevrolet Kingswood in der Lackierung Crown Sapphire (welch schöner Name) und Snowcrest White. Das rückwärtige Fenster ließ sich versenken, und es gab eine dritte Bank.*

*Rechts: Chevrolet Nomad in viertüriger Ausführung. Es dürfte nicht leicht gewesen sein, die ausladenden Heckflossen jenes Jahrgangs in das Kombiwagen-Design zu integrieren.*

Chrysler als drittgrößter Automobilhersteller Amerikas war stets mit Station Wagons vertreten, die in den frühen vierziger Jahren durch ihren Woody-Stil besonderes Aufsehen erregten. Aber es gab auch Limousinen, Coupés und Cabriolets in der Holz/Stahl-Ausführung; die Baureihe hieß »Town & Country«.

*Town & Country Coupé von 1948. Die Holzteile sind noch aus Mahagoni und Esche – anschließend stellte man die dunklen Teile aber wieder aus Stahlblech her; die Holzstruktur bestand aus einer aufgeklebten Folie.*

Die Marke De Soto wurde 1928 gegründet und gehörte zum Chrysler-Konzern. Als Besonderheit bot De Soto ab Ende 1948 eine viertürige Limousine mit umlegbarer Fond-Rückenlehne und durchgehender Ladefläche einschließlich Kofferraum an.

*1959er de Soto Firesweep Shopper Station Wagen. Die anderen Modelle trugen nicht weniger feurige Bezeichnungen, wie zum Beispiel Firedome und Fireflite.*

Dodge Kingsway
Commercial Utility

The foundation of the truly balanced 1954 Kingsway is the sturdy frame. Rigidity has been achieved with wide double channel side rails and sturdy, short body brackets. The result is that the car as a whole is better able to resist the twisting forces imposed on it by the road, and the car has a sound "feel." Frame and chassis are designed for balanced weight distribution. Rubber mountings effectively insulate the steering gear assembly from the frame to give easier steering, without annoying vibration. It has a balanced suspension system, with soft, wide rear springs angle-mounted, inside the frame. This permits the 1954 Kingsway to give you the full benefit of the famous Oriflow shock absorbers.

*Oben: 1955er Dodge Coronet Sierra Station. Der V8-Motor hatte 4,4 Liter Hubraum und 189 PS. Es gab auch 5,2- und 5,8-Liter-Aggregate.*

*Links: Aus einem Dodge-Prospekt von 1954. Abgebildet ist der zweitürige Kingsway Commercial Utility mit Sechszylindermotor.*

Die Dodge Brothers, seit 1914 im Automobilgeschäft, verkauften ihr Unternehmen 1928 an Chrysler. Schon 1924 gab es einen Dodge Woody als offenen Sechssitzer; der Dodge Westchester Suburban Wagon von 1937 war der erste Kombi in der auch später üblichen Station-Wagon-Art.

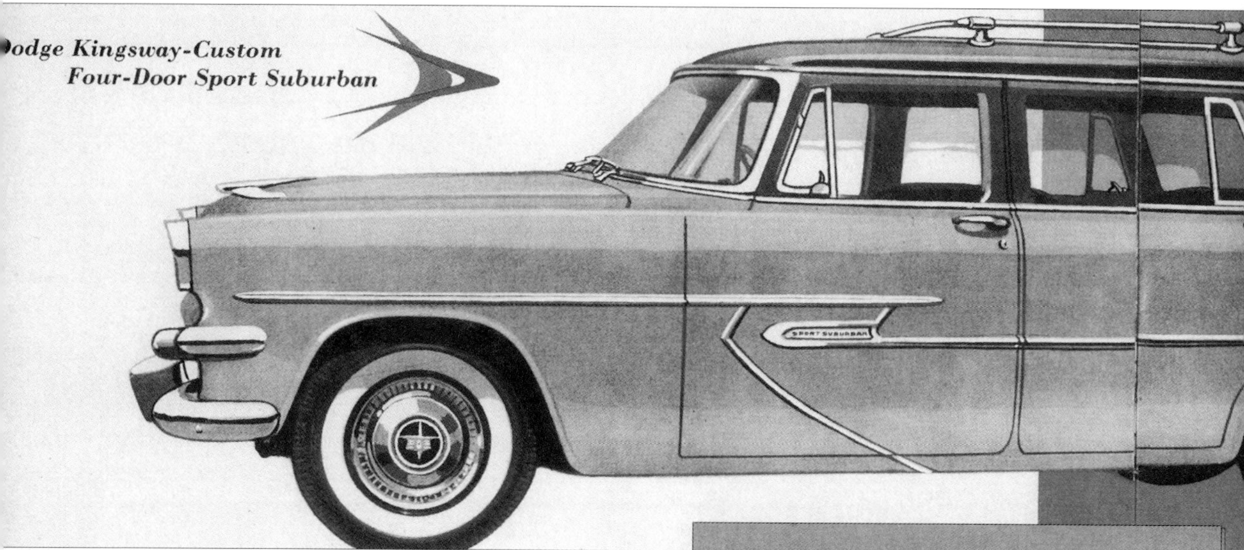


*Oben: Dodge Kingsway Custom Four-Door Sport Suburban – so die genaue Bezeichnung dieses Kombiwagens, der auf Wunsch mit einer dritten Sitzbank im Heck geliefert wurde.*

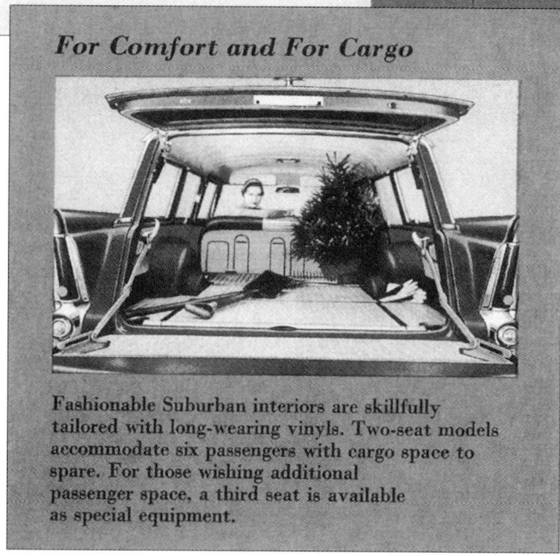

For Comfort and For Cargo

Fashionable Suburban interiors are skillfully tailored with long-wearing vinyls. Two-seat models accommodate six passengers with cargo space to spare. For those wishing additional passenger space, a third seat is available as special equipment.

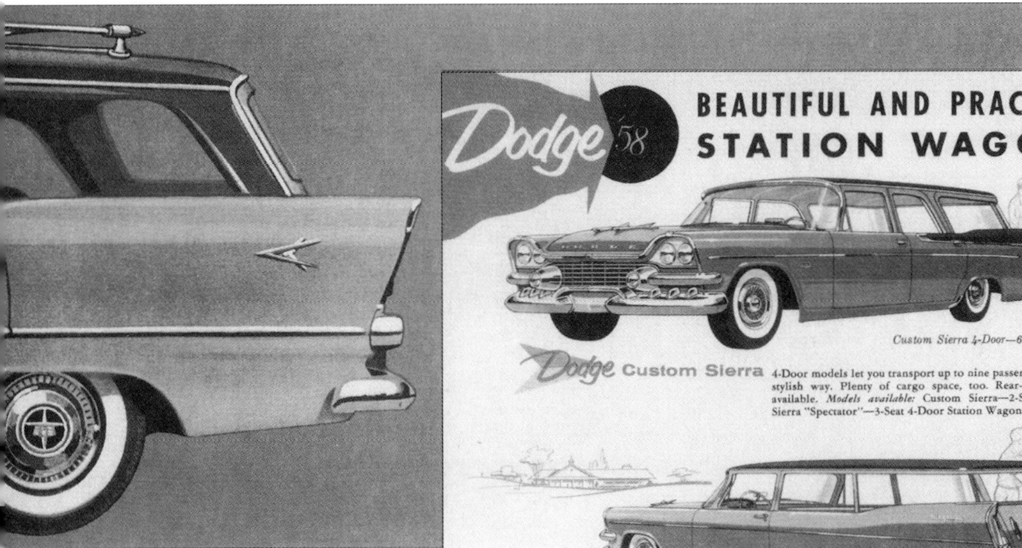

*Rechts: Dodge Kingsway Custom Sierra. Eine breite Skala von Modellen und Motoren wurde angeboten, wobei die Zweitürer die Ausnahme bildeten.*

Die Marke Edsel gehörte zum Ford-Konzern und bestand nur von 1958 bis 1960. Sie entwickelte sich zu einem Millionen-Flop, und die heute noch verbliebenen Fahrzeuge gehören zu den Raritäten auf dem Sammlermarkt.

*Im breiten Modellangebot gab es auch einige Edsel-Kombiwagen. Oben: Werbung für den 1959er Villager Station Wagon, den man – wie üblich – mit drei Sitzreihen erhalten konnte, weshalb man ihn durchaus als Neunsitzer bezeichnen durfte!*

*Fünf Station Wagon-Modelle waren zu bekommen, darunter auch ein Zweitürer namens Roundup. Den neunsitzigen Bermuda konnte man auch mit Woody-Dekor erhalten.*

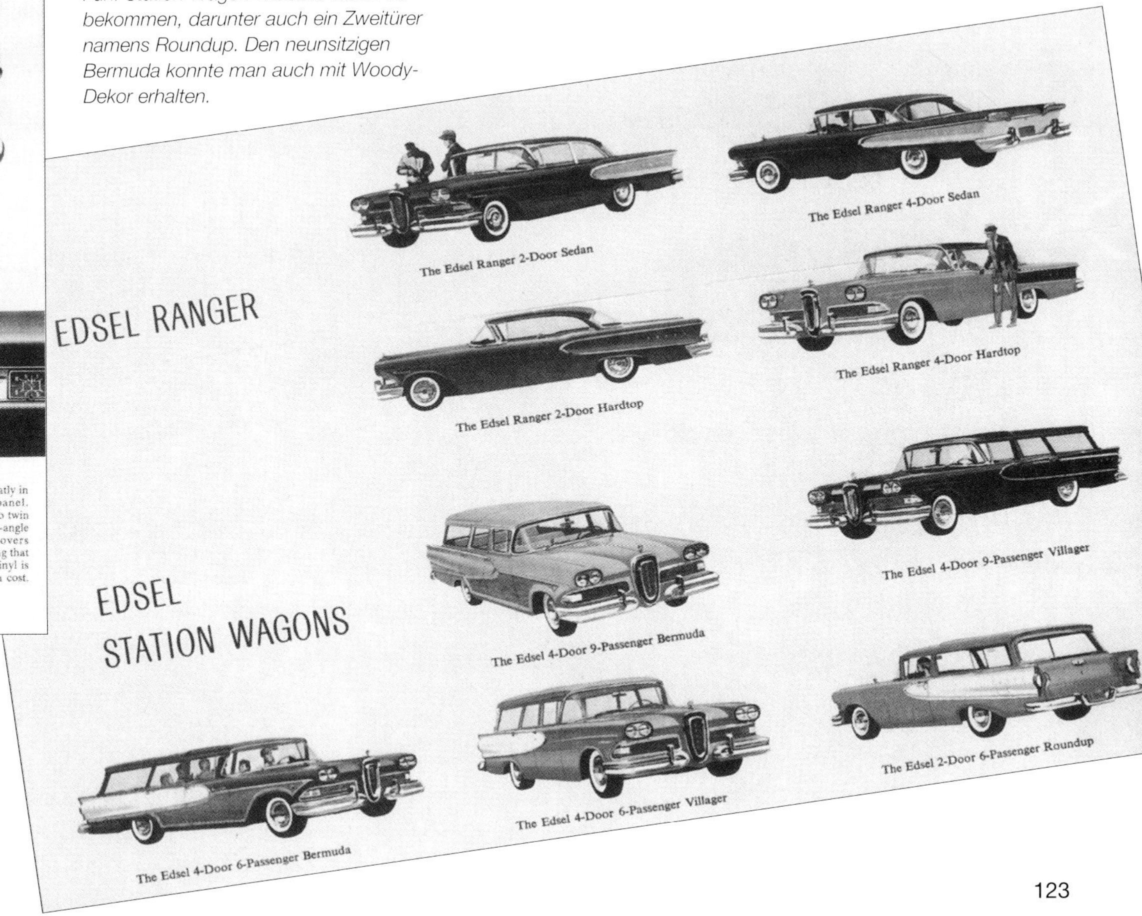

*In einem plakatgroßen Prospekt warb Ford für den 1958er Edsel, der eine ausgeklügelte Innenausstattung aufwies.*

*Angeblich war es die Wagenfront, die beim Kunden nicht »ankam« und den Edsel zu einem Außenseiter machte.*

Ford V8 de Luxe Station Wagon 1938. Die einen sprachen von einer primitiven Holzkiste, die anderen fanden solche Karosserien bildhübsch…

Die Ford Motor Company gehört zu den ältesten Automobilherstellern in den USA und brachte bereits vom T-Modell kombiwagenähnliche Versionen heraus; die Aufbauten lieferten allerdings Fremdfirmen. In den späten dreißiger Jahren gehörten Stahl/Holz-Karosserien dann ebenso zum Standard-Modellprogramm wie in den darauffolgenden zwei Jahrzehnten: Die Woodies von Ford waren sehr reizvoll.

*Ford Station Wagon 1939, Typ 92 A mit*
*60 PS starkem V8-Motor (2227 ccm).*

*1948er Ford Station Wagon. Wahlweise konnte man einen V8- oder einen 6-Zylindermotor bekommen. Dieser Wagen gehört einem schwedischen Sammler (Foto aus dem Motorhistorisk Magasin).*

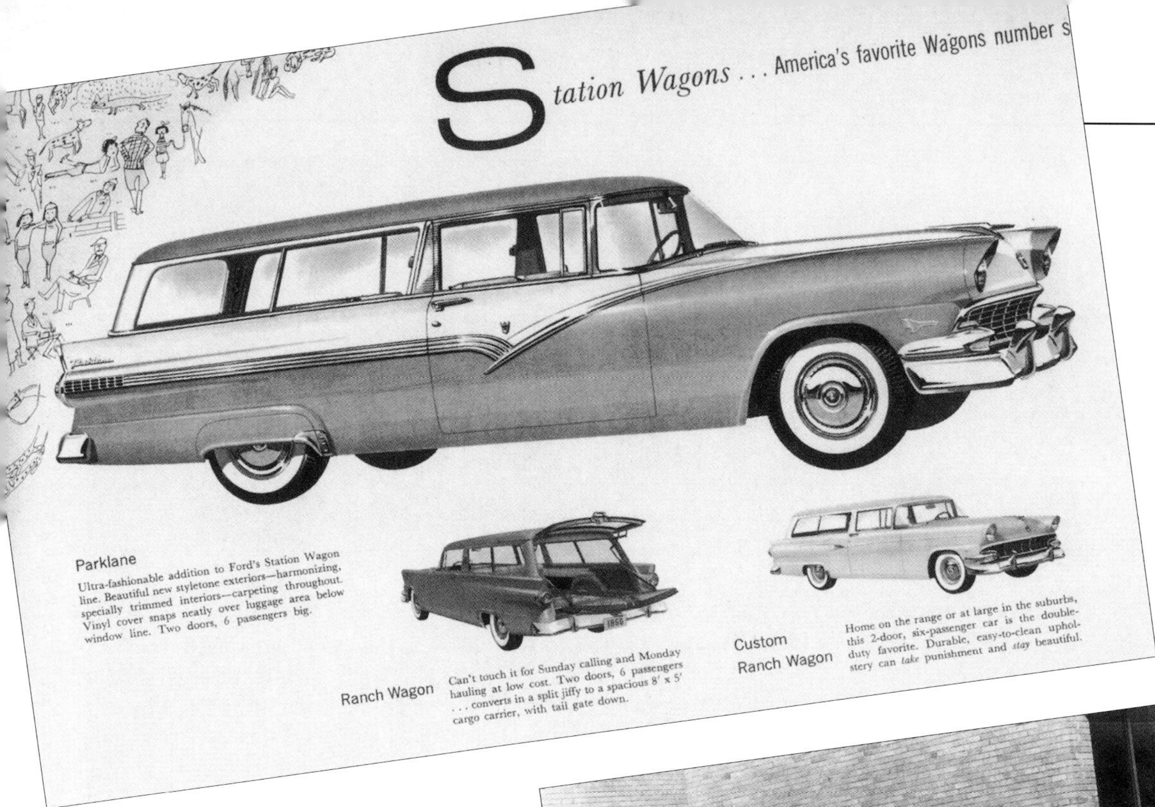


*Werbung für den Ford Parklane Station Wagon 1956, auch als »Ranch Wagon« oder »Country Sedan« bezeichnet und als Zwei- wie als Viertürer zu haben.*

*Rechts: Viertüriger Ford Country Sedan von 1953/54. Auf Wunsch waren bereits elektrische Fensterheber zu bekommen.*

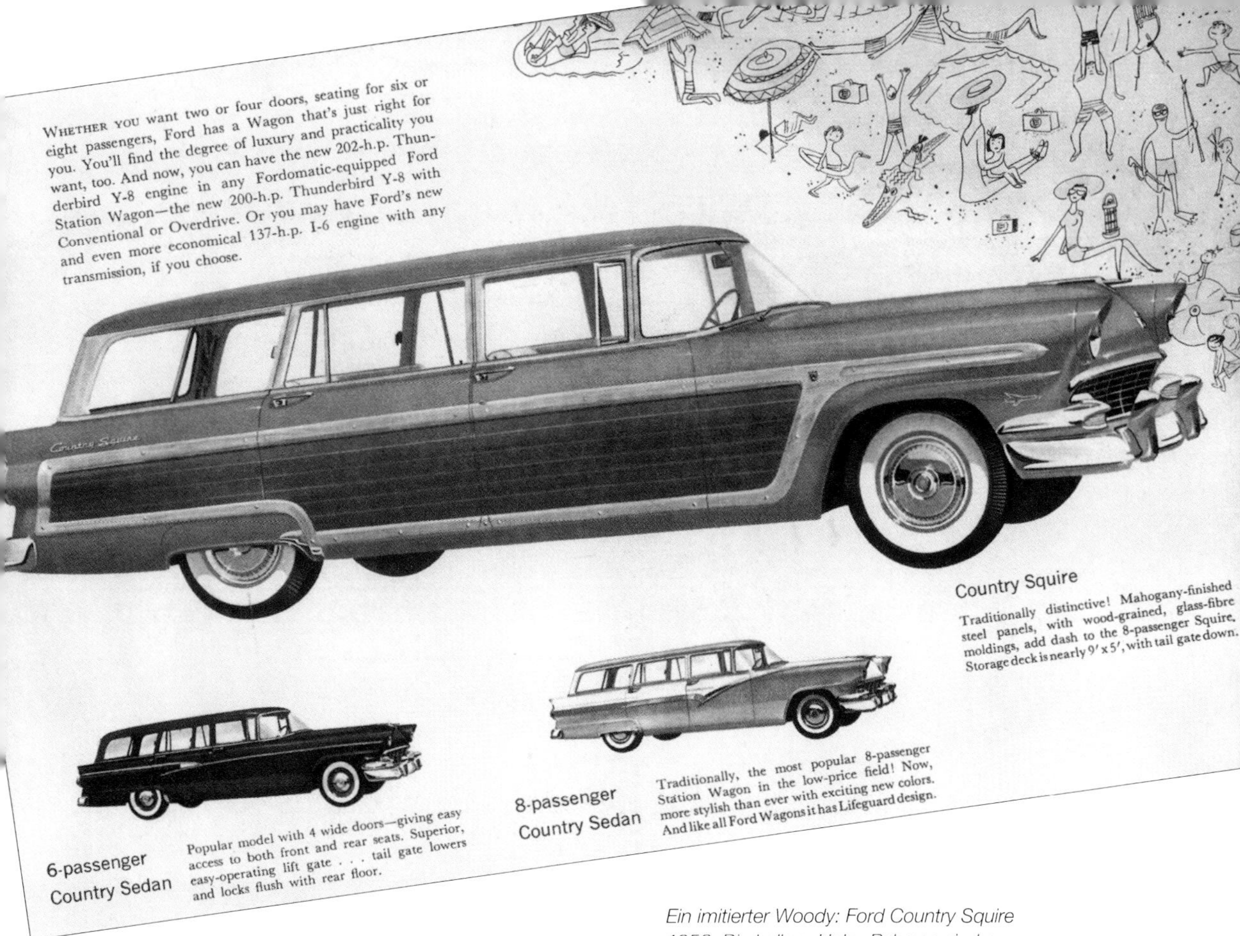


*Ein imitierter Woody: Ford Country Squire 1956. Die hellen »Holz«-Rahmen sind aus Kunststoff, die dunklen Flächen mit Holzmaser-Effekt lackiert. Immerhin: Ein eindrucksvoller Allroundwagen.*

*1957er Ford Ranch Wagon als »Traumboot mit einem Ozean an Platz«, wie es in diesem Prospekt heißt. Auch sei dieses Auto so preisgünstig, daß es ein idealer Zweitwagen sei. Diese 16seitige Werbeschrift war ausschließlich den Wagons gewidmet und gab ihre Geräumigkeit und den großen Komfort wieder. Elektrische Fensterheber und elektrisch verstellbare Vordersitze waren selbstverständlich. In konventioneller Bauweise war allerdings das Chassis gehalten: Es war ein Tiefbett-Kastenrahmen mit hinterer Starrachse.*

# Ranch Wagon

This 2-door, six-passenger dreamboat has oceans of space. Like all Ford wagons, it has a wider wrap-around liftgate with curved glass for better vision, better appearance and more cargo room. And while the Ranch Wagon plays a beautiful tune on your heart strings, it plays a lullaby on your purse! In fact, it's priced so low, that it makes an ideal second car.

From caller to hauler in 3 seconds flat! New tailgate has new two-stage latch release. Raise release part way and liftgate springs open. For '57, there's plenty of loadspace because the Stowaway seat folds down to form a long load floor. You'll appreciate the rugged upholstery (woven plastic or pleated vinyl) and the colorful ribbed linoleum flooring in the load area.

FORD „Station-Wagen"
– fünf neuentwickelte Ausführungen von bezwingender Schönheit
8-SITZER COUNTRY SEDAN
Viertüriger „Großraum"-Wagen mit reichlich Platz für 8 ausgewachsene Personen. Seine neue, betont niedrige Karosserie ist eine glückliche Kombination von bestechender Schönheit und vollendeter Technik: elegant, ruhig und sehr stabil.
Auch hier liegt FORD an der Spitze: Länger und niedriger denn je zuvor, bieten Ihnen diese Wagen jeden möglichen Luxus und außergewöhnliche Bequemlichkeit. Die Rücksitze lassen sich herunterklappen. Alle Bezüge sind aus robustem Plastik-Gewebe oder aus abwaschbarem Vinyl. Jeder dieser Wagen bietet reichlich Platz für großes Reisegepäck, für umfangreiches Sportzubehör, ja selbst für eine auf Wochen berechnete Campingausrüstung. – Auch die FORD-„Station-Wagen" verfügen über eine Frischluftzufuhr auf der ganzen Breite der Windschutzscheibe. Sie haben wiederum die Wahl zwischen den großen V8-Motoren und dem bekannten MILEAGE MAKER-Sechszylinder.
RANCH-WAGEN
Ein preislich besonders günstiges Modell mit zwei Türen. Viel Platz für 6 Personen mit reichlichem Gepäck oder für eine umfangreiche Ladung. Die Klapptüren lassen sich mit einem Griff nach oben und unten öffnen und schließen.
6-SITZER COUNTRY SEDAN
Hier verbindet sich alle Bequemlichkeit eines viertürigen Wagens mit großer Geräumigkeit: volle 2750 x 1525 mm mißt die Ladefläche bei heruntergeklappten Rücksitzen und geöffneter Hecktür.
DEL RIO RANCH-WAGEN
Ein eleganter Partner für geschäftliche Zwecke und für die Freizeit. Sein luxuriös ausgestattetes Innere bietet Platz für sechs Erwachsene mit vollzähligem Gepäck für die große Urlaubsreise.
8-SITZER COUNTRY SQUIRE
Dieser Wagen ist eine Besonderheit unter den amerikanischen „Station-Wagen" und sticht unter allen am meisten hervor! Seine geschmackvolle, eigenwillige Stahlkarosserie verleiht dem Fahrzeug eine ungewöhnliche Note.

Expect a whole new kind of excitement at your house when you arrive in your 1957 Ford Station Wagon. For you'll be the proud owner of a totally new kind of station wagon that's longer and lower than ever . . . 17 feet long . . . less than 5 feet high!
Expect to ride closer to the road than ever before. Thanks to the new "Inner Ford" you will. And yet you'll have full headroom . . . full road clearance. You'll ride silent, solid and secure.
Expect lots more "sizzle" and far fewer gas stops, too. You get
© 1956, Ford Motor Company, Dearborn, Michigan

*Links: Achtseitiger Prospekt für die 1957er Ford-Modelle, herausgegeben von der Ford Werke AG in Köln: »Der Wagen mit den Linien von morgen«. Den Station Wagon bot man in fünf verschiedenen Varianten an, wahlweise mit Sechs- oder Achtzylindermotor.*

*Unten: Amerikanische Dimensionen und ein »Glamour Wagon« mit Holz-Imitat und dicken Kunststoff-Nieten. In jedem Fall stellte der Woody, auch wenn er eigentlich keiner war, ein reizvolles Auto dar.*

*Oben: 1959 warb die Ford Motor Company mit der Medaillie, die man ihren Autos auf der Brüsseler Weltausstellung für »außergewöhnliche Formen und eine elegante Linie« zuerkannt hatte.*

*Rechts: Wie so oft, stellte man die Menschen in Relation zu den Autos besonders klein dar, so daß man den Eindruck hatte, es gingen nicht nur neun, sondern mindestens zwölf Leute in einen Ranch Wagon hinein...*

# WAGONS!

## RANCH WAGON

This spirited model is first choice with those who seek a 2-door wagon at the *lowest* price. Team it with Ford's Mileage Maker Six and you'll enjoy the *thriftiest* "wagon transportation" since the ox-drawn Prairie Schooner! Example: you change oil only every 4000 miles instead of the 1000 often recommended.

## FORDOR RANCH WAGON

It's so easy to load or unload passengers or payloads in this 4-door, 6-passenger wagon . . . yet it is priced with the lowest! And it's thrifty to maintain, too, with a new aluminized muffler that will normally last *twice* as long as conventional mufflers used on other wagons.

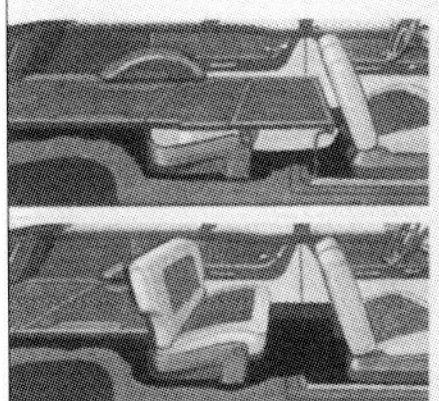

Fold the Stowaway seat down, as shown in the top picture, and you're all set to haul just about anything. Or, set the Stowaway seat up (below) and you're set to take 6 passengers in comfort!

## BEAUTY SURROUNDS YOU

Always known for the beauty of their station wagon interiors, Ford stylists have outdone themselves in these 6-passenger models (right). You'll find interiors that are as beautiful as they are practical . . . interiors that are color-keyed to be in perfect taste.

As for size, there's 92 cu. ft. of cargo space in every '59 wagon. They measure nearly *ten feet* from back of front seat to the end of the open tailgate (as seen in illustration below right). Truly, nothing measures up to America's *best-selling* wagons!

ALL SEATS FACE FORWARD IN FORD'S 9-PASSENGER WAGONS

*Argumente über Argumente, die für die Anschaffung eines 1959er Ford Wagon sprechen! Es gab auch zwei verschiedene Getriebeautomaten zur Wahl: Fordomatic und Cruise-O-Matic.*

*Am äußeren Ende des dargestellten Fahrzeugs wirken die Menschen besonders klein und das Auto um so größer. Ein alter Designertrick.*

THE NEWEST IN WAGONS FROM
Ford's six big station wagons for '59 bring you new space for the space age. And how your family can use it!
Here's the fresh new direction in station wagon design: hardtop styling. And how your neighbors will love it!
For '59 all Ford wagons are bigger . . . over 7 inches longer inside with up to 13 cu. ft. more cargo space riding on a 2-inch longer wheelbase. They have a still-larger glass area. And what's more, only Ford in the low-price field offers you the comfort of a
9-passenger, 3-seated wagon w
Both of Ford's standard engin
V-8—are designed to thrive
saves you up to 5¢ over pre
And new Diamond Lustre
won't have to wax it—ever!
With all their new proporti
hardtop styling . . . in loads
these 59 Ford wagons a mast
© 1958, Ford Motor Co., Dearborn, Mich.

*Der neunsitzige Ford Country Squire: Ein Auto für Zwerge? In künstlerischer Übertreibung wurde der Wagen überproportional lang dargestellt. Ford sprach von der »schönsten Proportion der Welt«.*

# Mercury

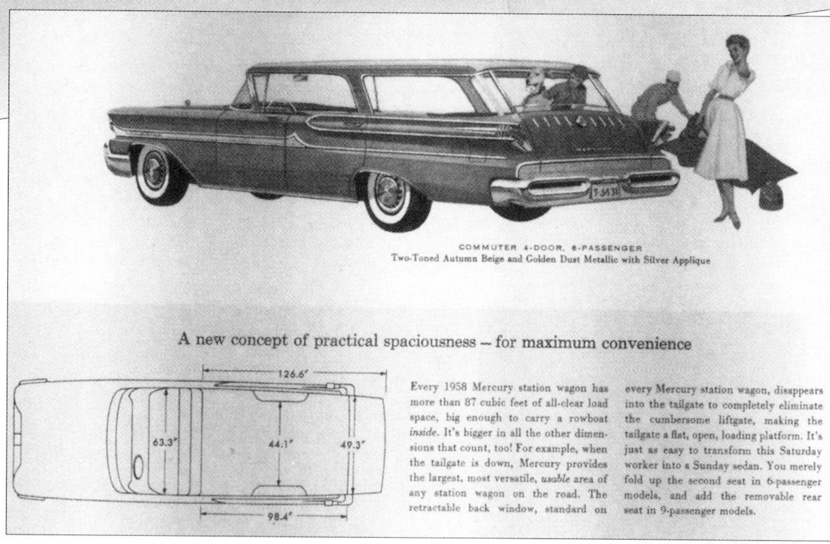


COMMUTER 4-DOOR, 6-PASSENGER
Two-Toned Autumn Beige and Golden Dust Metallic with Silver Applique

A new concept of practical spaciousness – for maximum convenience

Every 1958 Mercury station wagon has more than 87 cubic feet of all-clear load space, big enough to carry a rowboat *inside*. It's bigger in all the other dimensions that count, too! For example, when the tailgate is down, Mercury provides the largest, most versatile, *usable* area of any station wagon on the road. The retractable back window, standard on every Mercury station wagon, disappears into the tailgate to completely eliminate the cumbersome liftgate, making the tailgate a flat, open, loading platform. It's just as easy to transform this Saturday worker into a Sunday sedan. You merely fold up the second seat in 6-passenger models, and add the removable rear seat in 9-passenger models.

Die zum Ford-Konzern gehörende Automarke Mercury trat erstmals 1939 in Erscheinung. Kombiwagen waren von Anfang an im Programm, und ab 1946 auch im Woody-Stil. Seither gab es Mercury Station Wagons in zahlreichen Erscheinungsformen.

*Gegenüberliegende Seite: Werbung für den Mercury Commuter Modelljahrgang 1958. Raketenlook mit den Stilelementen eines Fachwerkhauses, aber mit rahmenlosen Scheiben in den Türen: Hardtop Styling!*

*Nur in ein paar Kleinigkeiten unterschieden sich zwei Prospekte, die 1958 den Mercury beschrieben. Oben eine Seite aus der revidierten Auflage. Es gab den Mercury Commuter zwei- und viertürig.*

*Mercury Voyager 1958. Das Hardtop-Styling läßt die Linien dieses Autos noch gestreckter erscheinen. Vier Motoren von 312 bis 400 PS standen zur Wahl.*

*Einer der wenigen Prospekte, die es vom Mercury des Jahrgangs 1957 auch in deutscher Sprache gab. Oben der Colony Park mit Holz-Imitat-Paneelen, in der Mitte der Voyager mit drei Sitzreihen, unten der zweitürige Voyager. Als besonders edle Variante gab es den viertürigen Commuter.*

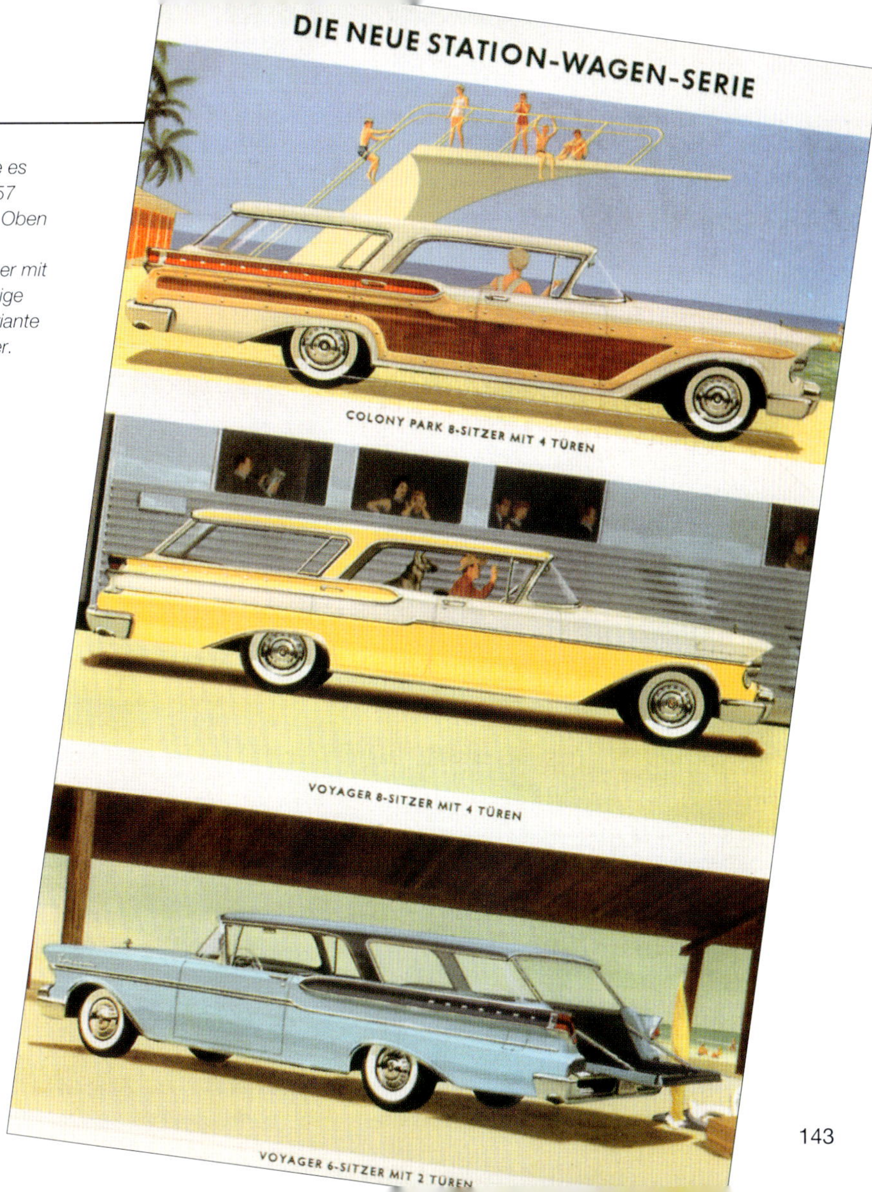

Two-Toned Bermuda Sand
and Marble White

Sherwood Green

*Amerikas lebendigster Luxuswagen: So beschrieb Mercury das 1959er Modellprogramm (links). Der Commuter zeichnete sich durch besonders große Glasflächen aus. Wie stets, gab es einen Vier- und einen Zweitürer. Links unten der viertürige Country Cruiser 1959. Von jetzt an nahmen die Autos an Breite und Länge nicht mehr zu...*

*Als ein sauberes dynamisches Design bezeichneten die Werbetexter die Linien des 1959er Mercury Country Cruiser im Woody-Stil: »Alles, was Sie sich auf Rädern nur wünschen können...«*

*Nash Rambler Suburban Station Wagon 1955.*

Bevor der Modellname Rambler zur eigenen Marke avancierte, benutzte ihn die Firma Nash für eine ihrer kleinen Baureihen. Der Rambler Custom Station Wagen war ein kompakter Kombiwagen, fast in europäischer Dimension.

*Nash Rambler Custom Station Wagon, aus einem Prospekt für das Modelljahr 1951. Nash zählte mit seinen selbsttragenden Ponton-Karosserien, die sogar die vorderen Räder umschlossen, zur Avantgarde. Dabei blieben die Abmessungen des Wagens in vernünftigen Grenzen.*

# La petite Rambler, maniable et sportive

## Dès son apparition en Suisse, »l'Américaine« ouverte la plus vendue

Deux nouveaux et élégants modèles sport complètent désormais la classique série Nash: Le cabriolet-limousine Rambler et le modèle Rambler »Station-Wagon« — deux véhicules de la classe luxueuse Custom, au prix le plus bas de toutes les voitures américaines de même classe et de même exécution. (Le Station-Wagon est livrable aussi en modèle »Super« avec équipement quelque peu simplifié).
Le cabriolet-limousine »Rambler« avec capote automatique allie la sécurité d'une limousine tout-acier à l'attrait particulier du cabriolet. Comme première voiture de ce genre avec carrosserie autoporteuse (construction indéformable »Airflyte«), elle possède — comparativement au cabriolet de construction ordinaire — l'avantage d'une absence persistante de bruits, tout en offrant aux passagers une augmentation considérable de la sécurité en cas d'accidents (les cadres de portes et de fenêtres forment avec la coque un ensemble des plus solides).

**Le cabriolet-limousine Rambler modèle Custom**

**D'un goût poussé jusque dans les moindres détails, le cabriolet-limousine Rambler et le modèle Rambler Custom »Station-Wagon« sont équipés en série d'une installation de climatisation à surpression, d'un appareil de radio et d'autres accessoires spéciaux d'une valeur globale d'environ deux mille francs.**

**Le »Station-Wagon« Rambler modèle Custom**

*Oldsmobile Station Wagon von 1946.*

Oldsmobile, 1897 gegründet und damit eine der ältesten Automobilmarken der USA, gehörte ebenfalls zu denen, die bereits 1939 einen Station Wagen im attraktiven Woody-Stil anboten.

*Rechts: 1966er Oldsmobile Dynamic 88 Fiesta Station. Interessanterweise stellt man die Wirtschaftlichkeit dieser Wagen groß heraus: Die Benzinpreise in den USA betrugen schließlich nur ein Drittel von dem, was man in Europa bezahlte.*

Packard, 1899 gegründet, ist eine der faszinierendsten Marken Amerikas und überlebte nur bis 1958. Kombiwagen – auch im Woody-Stil – bot Packard ab Modelljahrgang 1940 an. Die Holzpaneele waren anfangs durchaus echt, sahen aber doch wie nachträglich aufgesetzt aus...

*Oben: Packard Station Sedan 1948, ein Kombi mit starkem Personenwagen-Einschlag. Das Holz sieht so aus, als habe man es nachträglich angesetzt.*

*Rechts: Packard Eight 1948. Ein solcher Woody kostete damals 3161 Dollar – für das Cabriolet zahlte man 171 Dollar weniger.*

*Tad Burness, der Herausgeber der in den USA sehr beliebten »Car Spotters Guides«, betätigte sich auch als Illustrator und schuf eine Serie von Zeichnungen amerikanischer Klassiker. Im Vergleich zu den Werbe-Illustrationen jener Jahre hielt sich Tad Burness an die realen Proportionen der Autos.*

Plymouth, seit 1928 existierend und von Anfang an eine Marke im Chrysler-Konzern, stellte bereits 1939 Kombiwagen her.

*1939er Plymouth Station Wagon – ein zeitgenössisches Werbefoto.*

*Plymouth Plaza Station Wagon Jahrgang 1957/58. Bestellte man den Wagen als Neunsitzer, mußten die hinten Sitzenden nach rückwärts blicken.*

DE CHRYSLER EN PLYMOUTH STATIONS-WAGONS
PAREN HET AANGENAME MET HET NUTTIGE
WINDSOR TOWN & COUNTRY
CHRYSLER STATION-WAGONS
Windsor Town & Country, 4 deuren, 2 zetels
New Yorker Town & Country, 4 deuren, 3 zetels
De geheel stalen familiewagens kennen een stijgende populariteit. Toch zijn het de Chrysler Town and Country en de Plymouth Suburban en Utility die de toon aangeven. Ze verenigen op perfekte wijze het luxueuse met het nuttige. De Chrysler New Yorker en Plymouth Suburban beschikken over 3 zetels om 9 personen in het grootst mogelijke komfort te vervoeren. Een demonstratie zal U doen ontdekken welke diensten deze wagens U kunnen bewijzen.
PLAZA COMMERCIAL UTILITY
1958
PLYMOUTH SUBURBAN EN UTILITY MODELLEN:
Plaza Commercial en Utility, 6 cyl. en V-8
Savoy Commercial Utility, 6 cyl. en V-8
Savoy Custom Suburban, 6 cyl. en V-8, 2 zetels
Savoy Sustom Suburban, 6 cyl., 3 zetels
Belvedere Sport Suburban, 6 cyl. en V-8, 2 zetels
Belvedere Sport Suburban, 6 cyl. en V-8, 3 zetels
MoPAR
De in deze folder voorkomende wagens zijn afgebeeld met enkele accessoires, die mits gering supplement kunnen geleverd worden. merken, uitrustingen en accessoires kunnen gewijzigd worden zonder voorbericht en zonder verplichting voor de Chrysler Corp., Export Division en/of de Chrysler Western Hemisphere Corp.

*Oben: 1959er Plymouth Suburban.*

*1958er Plymouth Station Wagon. Das Angebot umfaßte Versionen als Plaza, Savoy und Belvedere. Die Typen Windsor und New Yorker trugen – wie einst bei Chrysler – die Bezeichnung »Town & Country«.*

STATION WAGONS

*Sport Suburban V-8 4-door (6 or 9 passenger)*

*Custom Suburban 4-door (6 pass V-8 or 6, (9 passenger) V-8 only*

*Custom Suburban V-8 2-door (6 passenger)*

*DeLuxe Suburban 4-door (6 passenger) V-8 or 6*

*DeLuxe Suburban 2-door (6 passenger) V-8 or 6*

SAVOY

*4-door Sedan, V-8 or 6*

*2-door Sedan, V-8 or 6*

*Business Coupe 6*

**COOLING SYSTEM** Coolant capacity 318" V-8 engines 20 quarts; with heater 21 quarts; 361" V-8 engine 16 quarts; with heater 17 quarts; 6-cylinder 13 quarts; with heater 14 quarts. High-pressure (14 psi) system. 4-blade wing-tipped fan. Full-length water jacketing. Ball bearing water pump with permanent seal. Series-flow porting on 361" engine. 180° thermostat standard on all.

**SPECIFIC BODY DIMENSIONS**

| | 4-door Hardtop | 4-door Sedan | Club Sedan | 2-door Hardtop | Convert-ible |
|---|---|---|---|---|---|
| Front hiproom | 63.0" | 63.0" | 63.0" | 63.0" | 63.0" |
| Rear hiproom | 62.7" | 62.7" | 62.7" | 56.0" | 56.0" |
| Front shoulder room | 60.5" | 60.5" | 61.0" | 61.0" | 61.0" |
| Rear shoulder room | 60.4" | 60.4" | 60.4" | 60.5" | 55.0" |
| Front legroom | 45.5" | 45.5" | 45.5" | 45.5" | 45.5" |
| Rear legroom | 39.0" | 41.5" | 41.5" | 36.5" | 36.5" |
| Front headroom | 34.4" | 35.7" | 35.7" | 33.7" | 35.0" |
| Rear headroom | 34.5" | 34.2" | 34.2" | 34.0" | 35.2" |
| Car height—loaded | 54.6" | 56.6" | 56.6" | 54.0" | 54.8" |

**ACCESSORIES.** Many accessories have been engineered especially for the new Plymouth to make your driving easier and riding more enjoyable. Ask your dealer about these:

Push-Button PowerFlite (2-speed fully automatic transmission) • Push-Button TorqueFlite (3-speed fully automatic transmission) • Sure-Grip Differential • Overdrive • Safety-Sure Power Brakes • Constant Control Full-Time Power Steering • Power Windows • Airtemp Air Conditioner • Fresh-Aire Heating-Defrosting System • Instant gas heating system • Custom Push-Button Radio • Rear-Seat Speaker • Sweep-Second Electric Clock (standard on Fury models) • Chrome Bumper Wing-Guards • Rubber Bumper Protectors • Wheel Covers • Safety-Padded Instrument Panel • Safety Padded Sun Visors • Concealed Locked Luggage Compartment (6-passenger station wagons) • Power Tailgate Window • Solex-Tinted Glass • Undercoating • Whitewall Tires • Mirror-Matic Rear View Mirror • Remote Control Rear View Mirror • Prism Rear View Mirror • Automatic Beam Changer • Constant Level Torsion-Aire • Sport Deck • Swivel Seats.

For specifications of the new Plymouth STATION WAGONS, ask your dealer for the special STATION WAGON BOOKLET.

Cars illustrated are shown with items of optional equipment available at extra cost.

The policy of Plymouth Division of Chrysler Corporation is one of continual improvement in design and manufacture wherever possible to assure a still finer car. Hence, specifications, equipment and prices are subject to change without notice.

*Mit dem Modelljahrgang 1959 gab es bei Plymouth einen Höhepunkt der Heckflossen-Eskalation. Bei den Station Wagons wirkten die Flossen ohnedies absurd. Links das gesamte Plymouth-Programm 1959, das immerhin fünf Kombiwagen-Modelle umfaßte.*

Pontiac entstand 1926 als eine Gründung von General Motors und galt als preisgünstige Alternative zum Oakland. Anfangs gab es nur Sechszylinder, ab 1932 aber auch Modelle mit V8-Motor. Ab Ende 1938 bot Pontiac auch Kombiwagen an.

*Pontiac Station Wagons 1949. Es gab den Woody mit »echtem« Holz und auch mit einem Holz-Imitat!*

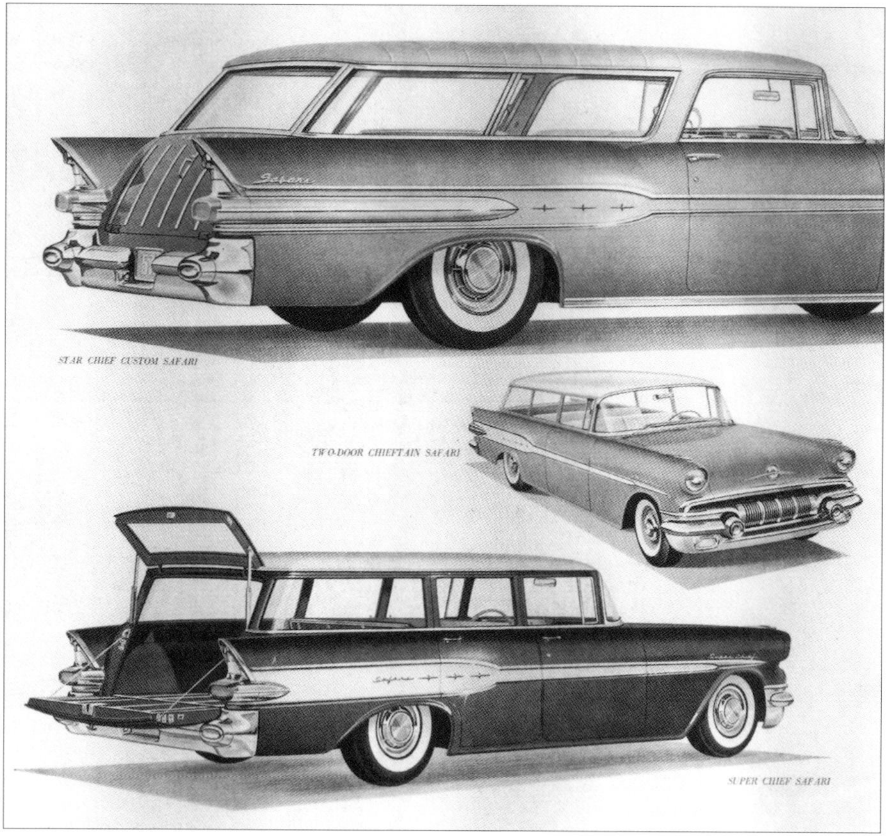


*Pontiac Safari 1957, »America's Number One Road Car«. Den Chieftain Safari gab es zwei- und viertürig.*

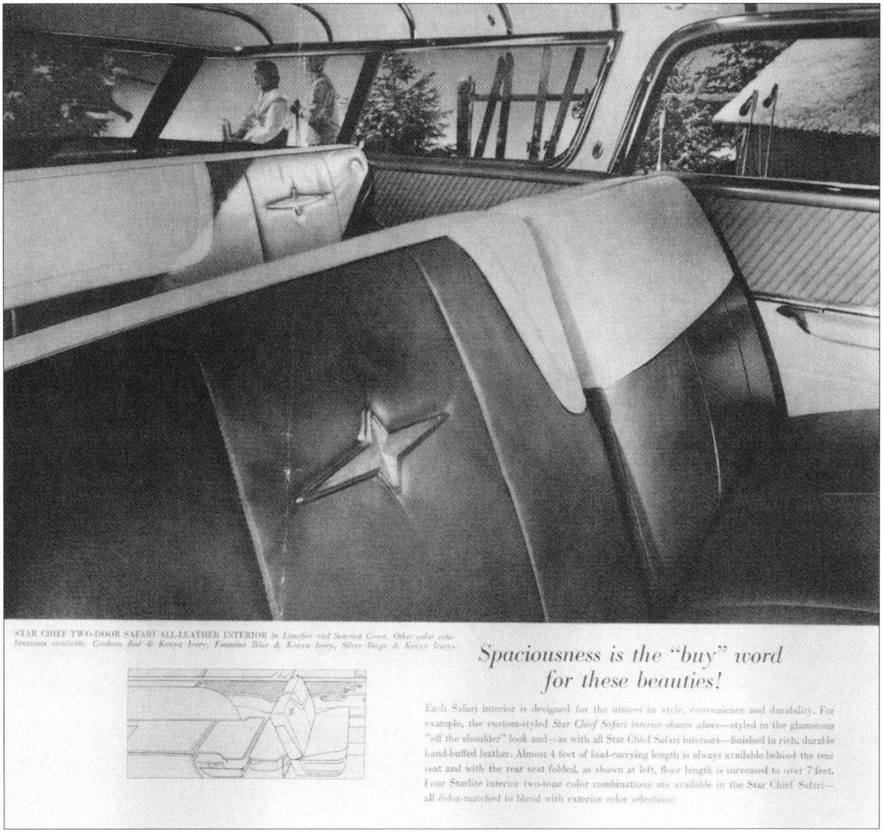


*Pontiac Safari Station Wagon 1957.
Einzelsitze waren in den USA nicht üblich,
doch die Bezüge erhielt man auf Wunsch
aus echtem Leder.*

*Nach dem Motto: Je länger, je lieber!*
*Pontiac Bonneville Custom Safari 1959.*

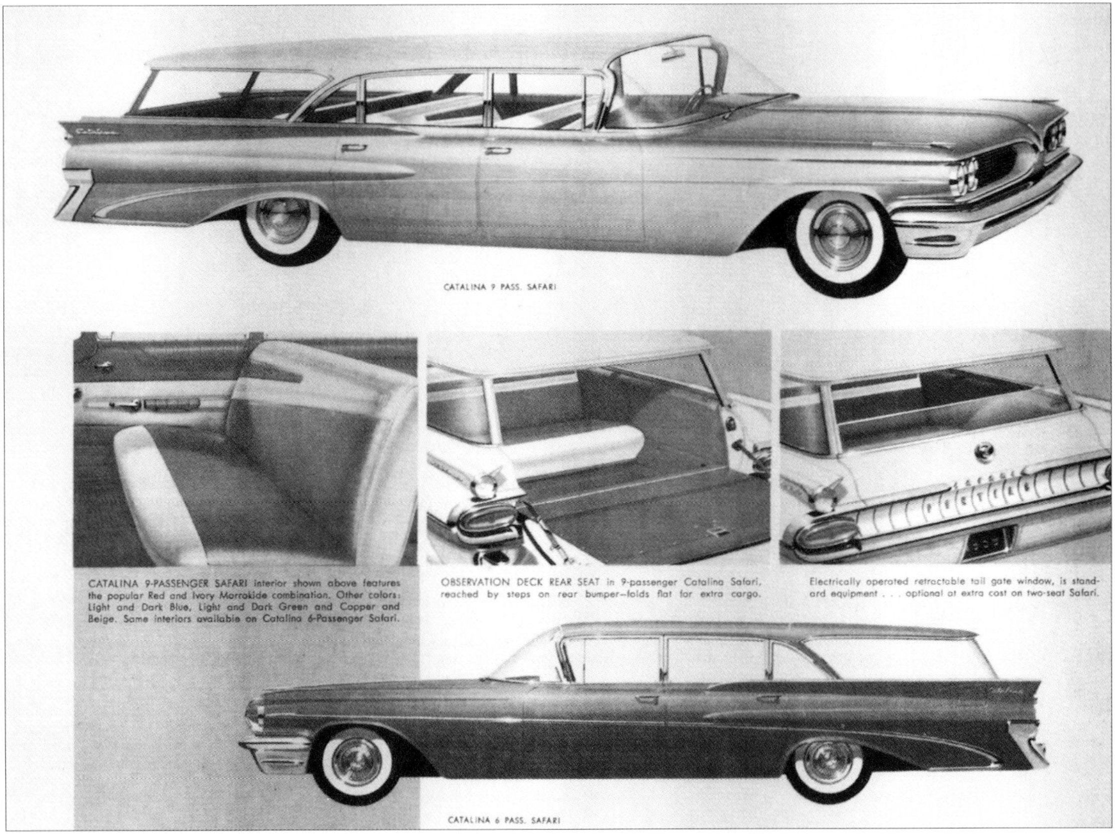


*Pontiac Catalina Safari mit drei Sitzreihen und elektrisch versenkbarer Heckscheibe.*

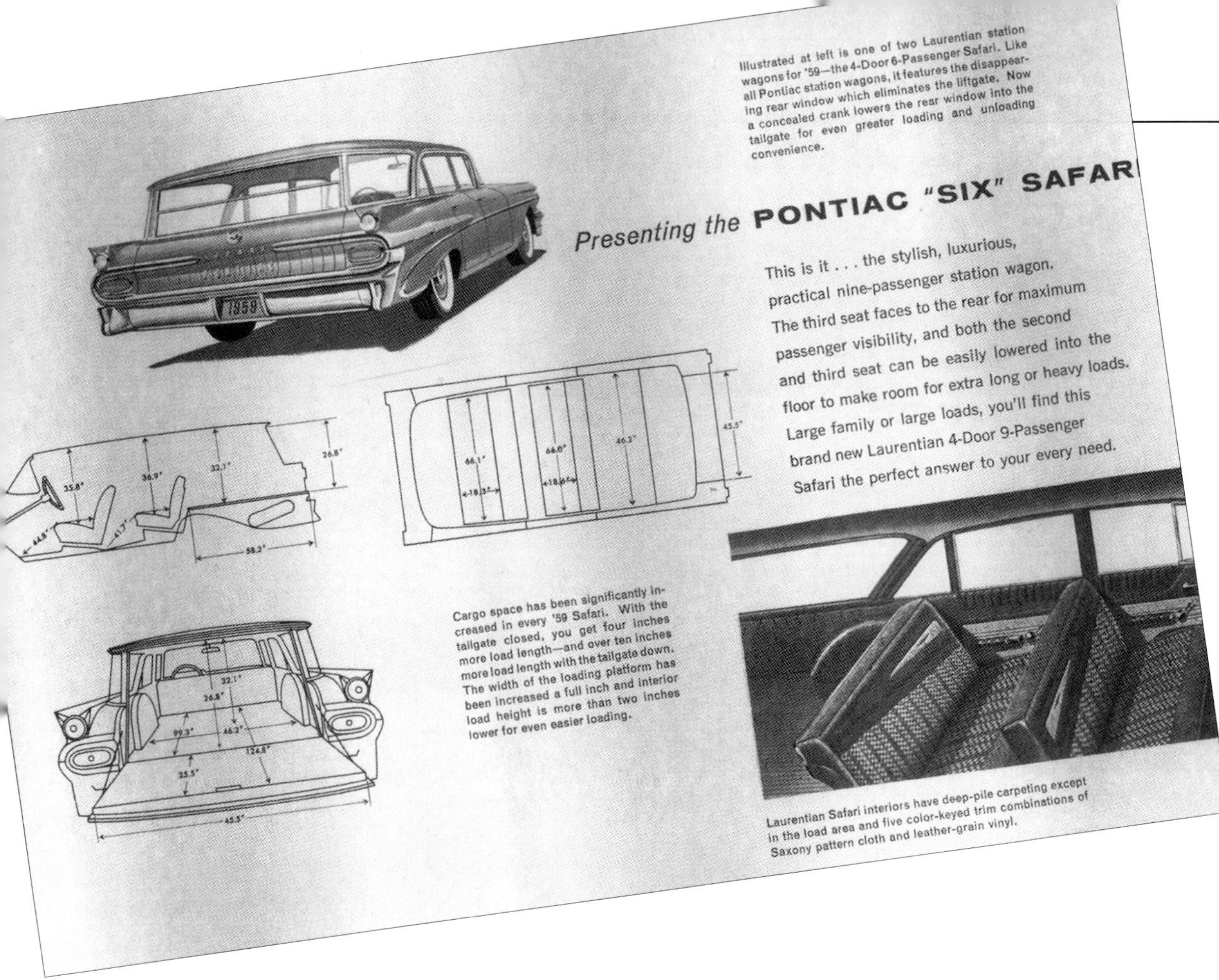


*Prospekt für den 1959er Pontiac Six Laurentian Safari, als Sechs- oder Neunsitzer erhältlich.*

*In der üblichen Übertreibung ließ man den ohnedies sehr großen Wagen noch ein wenig größer erscheinen, indem man kleine Kinder im Fond Platz nehmen ließ.*

*Werbung für den 1959er Pontiac Bonneville Custom Safari Station Wagon mit 300-PS-Motor.*

*Unten: In dieser Darstellung wirkt das Auto fast doppelt so lang wie in Wirklichkeit. Diese Wagen hätten einen hydraulischen Niveau-Ausgleich vertragen können...*

*Pontiac Catalina Safari 1960. Die Liste der Extras reichte vom Doppelrohr-Auspuff bis zu elektrisch verstellbaren Sitzen; Außenspiegel oder eine Scheibenwaschanlage waren im Serienumfang eines solchen Wagens damals allerdings noch nicht enthalten.*

American Motors verwendete von 1958 bis 1970 den Namen Rambler als eine eigene Markenbezeichnung, nachdem es vorher bei Nash einige Modelle dieses Namens gegeben hatte.

*Rambler Ambassador Cross Country Station Wagon in Hardtop-Bauart von 1960.*

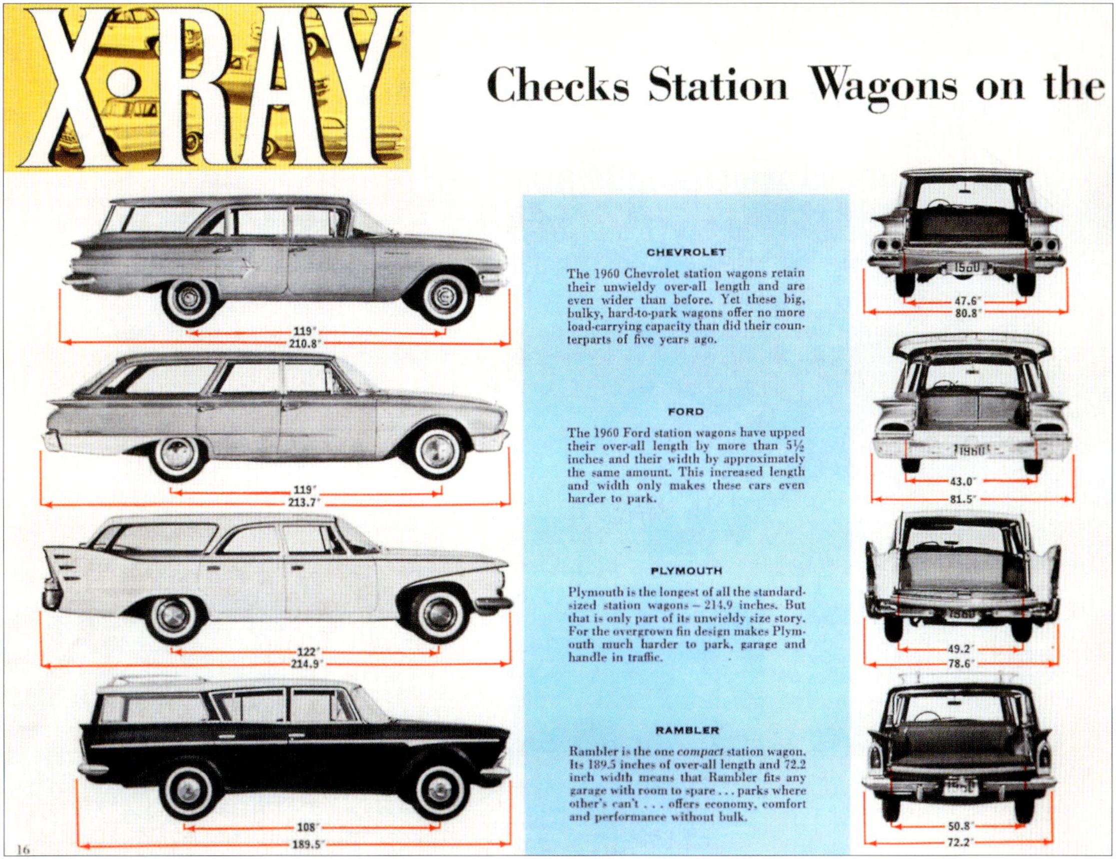


*Vergleichende Werbung war in den USA kein Tabu. Hier vergleicht man den Rambler Station Wagen mit Kombiwagen der Marken Chevrolet, Ford und Plymouth, um zu beweisen, daß der Kleinste der Größte ist!*

*Aufwendig gestalteter Prospekt für Rambler Automobile Jahrgang 1959. Die American Motors Corporation plädierte an die Vernunft der Autokäufer und warb mit Darstellungen ihrer Fahrzeuge ohne jegliche Übertreibung. Die Autos waren kompakt und von vernünftiger Dimension, wenn auch nicht ganz frei von modischen Accessoires, wie sie in den USA üblich waren.*

*Rechts: Ein Hauch von Woody findet sich im Dekor des Rambler Custom 4-Door Cross Country Station Wagon wieder. American Motors gab an, ihre Kombis seien die »beliebtesten« in den Staaten.*

*Studebaker Scotsman Station Wagon 1958.*

Studebaker, im Automobilgeschäft seit 1902, hatte stets ein großes Lastwagen-Programm anzubieten, folgte dem Kombiwagen-Trend aber erst verhältnismäßig spät. 1965 verschwand der Name Studebaker von der Szene.

*Links: Studebaker Pinehurst Station Wagon als Zweitürer – im Vergleich zur Konkurrenz eher dezent und im Dekor fast bescheiden.*

*Rechts: 1956er Studebaker-Modelle, von denen behauptet wurde, sie »strahlen hohes handwerkliches Können aus, das Ihnen die Gewähr gibt, genau den Station Wagon zu bekommen, den Sie brauchen!« In der Tat: Die Verarbeitung des Studebaker war stets ausgezeichnet.*

WAGONS

In every one of these brilliantly styled, highly practical Station Wagons by Studebaker you get the handling of a smart sedan with all the multi-purpose features of a station wagon. Specially engineered rear springing does the trick! Whichever one you choose has a special *flair* for doing your family proud–whether hauling home antiques from a backroads auction or parked in front of the country club!

PARKVIEW

In this high-spirited, sturdy and distinctive station wagon, you have an exciting choice of handsome, easy-to-clean and almost wearproof vinyl upholstery in striking patterns and colors. It's powered by the 170-hp Studebaker Sweepstakes 259 V8 for dazzling performance on any road.

PELHAM

Like all big, new Studebaker Station Wagons, the Pelham features new rotary Safe-lock doors; self-energizing *and* self-centering brakes; and specially engineered rear springing to make light-load riding soft and relaxing! Powered by Sweepstakes 185 Six.

POWER ASSISTS

POWER STEERING

What could be easier than having 75 per cent of the effort it takes to park supplied for you? A mere two full turns and your wheels turn from center to full right or full left. Yet you always have normal "road-feel."

POWER BRAKES

Just a slight touch on the brake pedal stops you evenly, safely – with up to 40 per cent less effort. And since *all* Studebaker brakes are self-energizing and self-centering, you never sense any sidesway or wheel pull.

POWER WINDOWS

Whether a 2-door or 4-door, you can have these electrically-controlled muscle-savers! Each window has its own pushbutton. And a master control lets the driver control all windows.

POWER SEAT

Your finger selects the exact position you want–for easier, safer, more relaxed driving. A single button at the bottom of the front seat electrically adjusts the seat precisely as you want it. Custom-fits car to every driver and family!

POWER AIR CONDITIONER

Even on the muggiest days, it's cool in your Studebaker. Outside air is dehumidified, filtered, cooled, and recirculated. One control knob for temperature; other two, circulation.

*Ende der Bescheidenheit: Nach der Fusion mit Curtiss-Wright gab es einen Ruck quer durch die Modellpalette. 1957 folgte man der von der Konkurrenz eingeschlagenen Linie und schmückte seine Commander-Baureihe ebenfalls mit überbordendem Chrom und Heckflossen im Übermaß. Viertüriger Station in der Sechszylinder-Ausführung von 1958, aus einem niederländischen Sammelprospekt.*

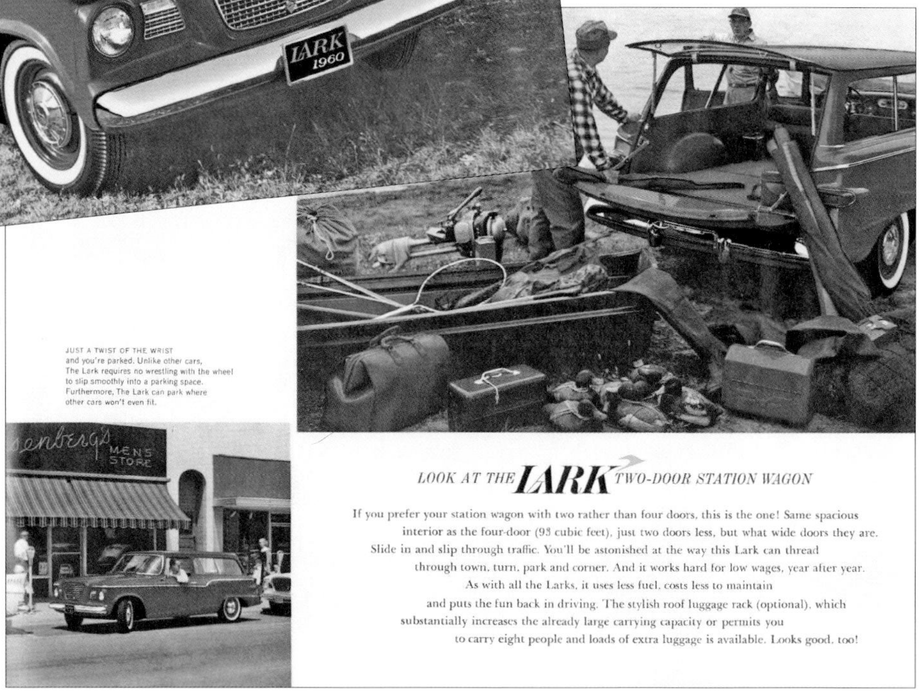


*Der Lark erschien für das Modelljahr 1959 und löste die bisherigen Sechs- und Achtzylinder-Typen Commander und Champion ab. Er überraschte durch seine schlichte, sehr kompakte Karosserie, die um gut 70 Zentimeter kürzer gehalten war als bisher. In das Modelljahr 1961 flog die »Lerche«(Lark) mit modernen, kopfgesteuerten Sechszylindern.*

Die Marke Willys-Overland wurde vor allem durch den US-Army-Jeep berühmt, kam 1953 zum Kaiser-Konzern, später zu Chrysler. Inzwischen war die Marke Jeep zu einem eigenständigen Fabrikat avanciert, das über Renault zu Chrysler gelangte. Von den späten vierziger Jahren an gab es vom zivilen Jeep stets Station Wagon-Versionen, teils als echte Woodies, meist aber als zweitürige Kombiwagen. Häufig wurden sie mit Zweirad-, seltener mit Vierradantrieb ausgeliefert.

*Oben: 1946 in Deutschland zum Woody-Kombiwagen umgebauter US-Jeep.*

*Unten: Formgestanzter Prospekt für den Willys Jeep Station Wagon von 1951.*

# Die europäische Konkurrenz

Kombiwagen sind keine amerikanische Erfindung, aber sie fanden in den USA eine weitere Verbreitung als bei uns. Und als man drüben mit dem Bau der »Woodies« begann, gab es sehr bald eine Reihe von Karosserieherstellern, die dem Trend folgten und auch in Europa die Stahl/Holz-Kombination populär machten, wie einige Beispiele auf den folgenden Seiten zeigen.

*Rechts: 1952er Alfa Romeo 1900 mit einer von der Firma Viotti angefertigten Woody-Karosserie.*

*Rechts: Auch der geländegängige Alfa Romeo Matta der fünfziger Jahre war mit Woody-ähnlicher Karosserie zu bekommen, doch handelte es sich um eine Ausführung in Stahlblech.*

*Borgward Hansa 1500 von 1950, einer der ersten neuen Kombiwagen in Pontonform in Deutschland nach dem Kriege.*

*Kleiner Kombi aus der ehemaligen DDR: Zwickau P 70 von 1957. In der Form zweckmäßig und ansprechend.*

*Die vom Fiat 500 C abgeleitete Giardinera war einer der reizvollsten Kombiwagen der kleinen Wagenklasse. Der Zweisitzer »Topolino« war hierdurch zum Viersitzer geworden und hatte vor allem eine breite Tür im Heck, die zum Be- und Entladen sehr praktisch war.*

*Englischer Ford V8, Modell Pilot, von 1948. Ein Echtholz-Woody, den es in nur sehr wenigen Exemplaren gab.*

*Rechts: Ford Köln war schon früh mit Kombiwagen zur Stelle, so auch mit diesem 17M Turnier mit 550 kg Zuladung, in dieser Ausführung 1959-60 gebaut.*

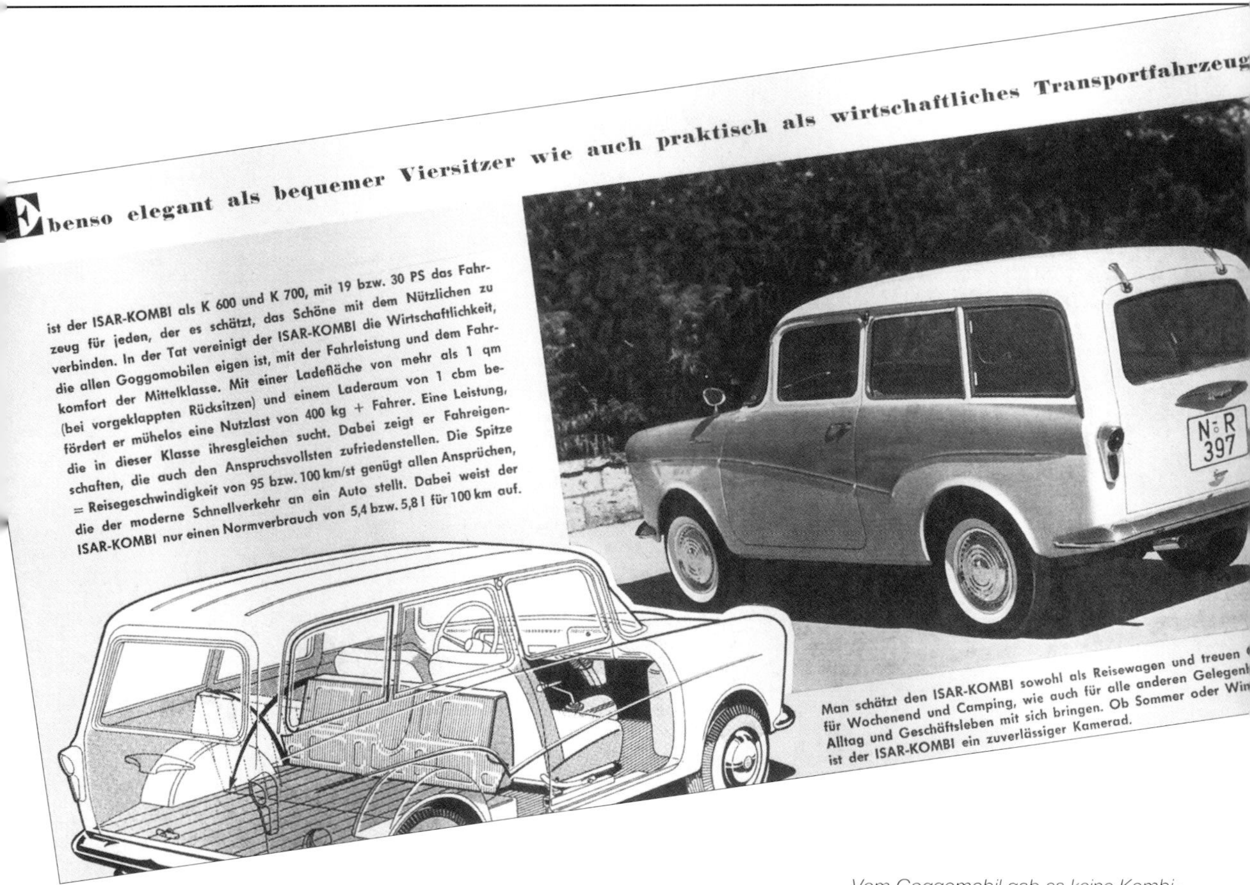


*Vom Goggomobil gab es keine Kombiwagen-Version, wohl aber vom Glas Isar K 600/700. Zuladung: »Fahrer plus 400 Kilogramm«, wie es in der Werbung hieß.*

*Links: Goliath Dreirad-Lieferwagen als Woody – kein ungewöhnlicher Anblick auf den Straßen in jener Zeit.*

*Den vierrädrigen Goliath Kleinlaster mit einem solchen Aufbau sah man schon seltener. Viele Exemplare dürften sicher nicht entstanden sein.*

*Ein Einzelstück blieb dieser in Belgien gebaute Imperia von 1928. Ein sehr früher Kombiwagen, der in den sechziger Jahrern als Reklamefahrzeug eingesetzt wurde.*

*Rechts: Große Popularität erlangten die im Woody-Stil aufgebauten Kombis auf Basis der Morris-Modelle Isis, Oxford und Minor. Sie trugen den Namen »Traveller«. Der Prospekt stammt von Ende 1956.*

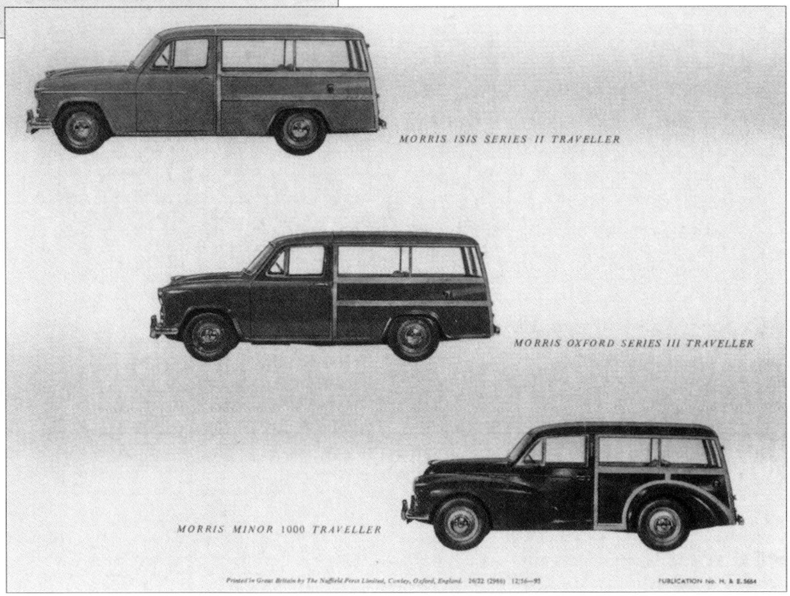

Der Morris Minor Traveller war ein echter Woody, und es gibt heute noch Spezialisten, die für Nachschub der Holzteile zur Restaurierung solcher Autos sorgen.

Der oben abgebildete Prospekt ist von 1954; links ein Faltblatt von 1960.

*Ein in Belgien gebauter Scaldia Kombi von 1965 – ein Montagefahrzeug, das in seinem Heimatland besser als »Moskwitsch« bekannt war...*

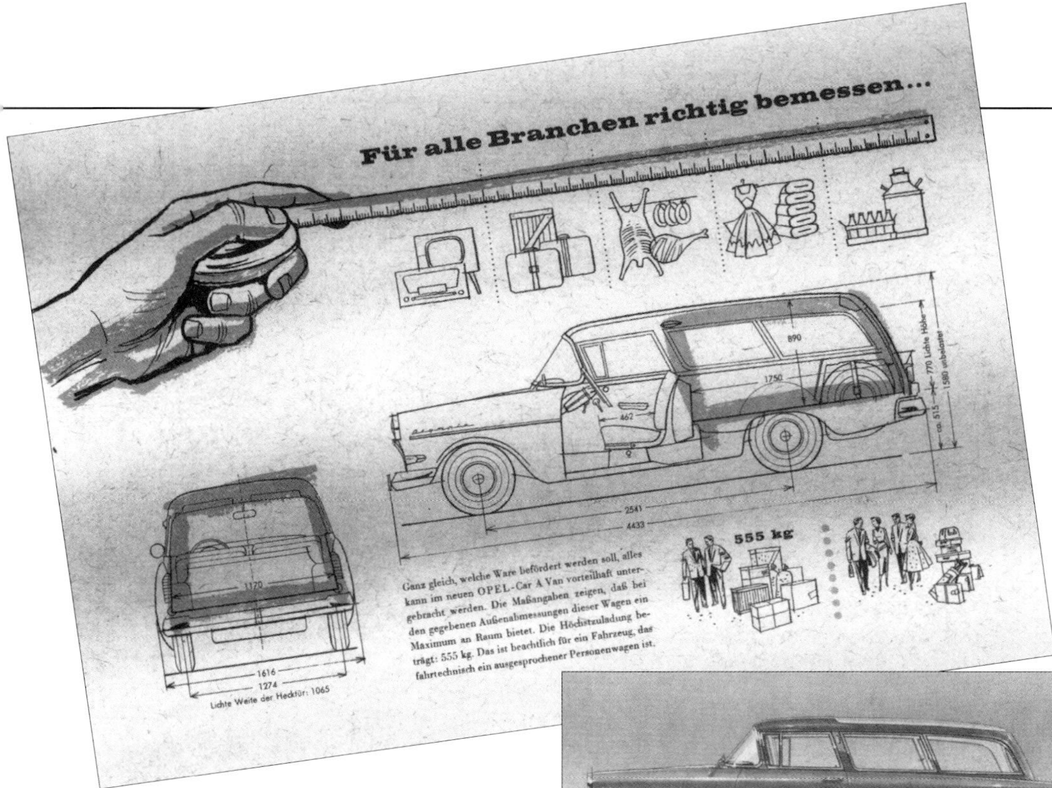

*Einer der berühmtesten Kombiwagen deutscher Herkunft war der Opel Caravan (auch Car-A-Van geschrieben). Oben eine Werbeschrift von 1957, rechts ein großformatiger Farbprospekt aus der gleichen Zeit.*

*Zweimal Rolls-Royce: Oben ein von Hooper hergestellter Aufbau auf dem Chassis eines 20/25 von 1929, rechts ein Silver Ghost 1924 als »Shooting Brake«; dieser Aufbau wurde indessen erst 1954 angefertigt.*

*Links: Skoda 1102 von 1950 als Woody-Kombi, den man als »Stations- oder Sanitätswagen« anpries.*

*Rechts: Ein Skoda Octavia Kombi von 1962.*

*Tempo Dreirad-Lieferwagen konnte man in den frühen fünfziger Jahren ebenfalls als Woodies bekommen. Sie waren echte Kombis mit einer zweiten, umklappbaren Sitzbank.*

*Die Kombiwagen-Ausführung des »Buckel-Volvo« hieß PV 445; man sah dieses Auto in den fünfziger Jahren in Skandinavien sehr häufig. Ein Prospekt von Ende 1954.*

*Wartburg »Camping«, ein etwas eigentümlich aussehender Kombi aus Eisenach mit Panorama-Dachfenstern und einem Rolldach darüber.*